The Besterman World Bibliographies

The Besterman World Bibliographies

Biological Sciences

A BIBLIOGRAPHY OF
BIBLIOGRAPHIES

By Theodore Besterman

TOTOWA, N. J.

Rowman and Littlefield

1971

Published by
Rowman and Littlefield
A Division of
Littlefield, Adams & Co.
81 Adams Drive
Totowa, N. J. 07512

★

Copyright © 1939, 1947, 1965, 1971
by Theodore Besterman
Printed in the United States of America

★

Typography by George Hornby
and Theodore Besterman

ISBN 0-87471-055-3

Contents

Preface

I have explained in the Introduction to the successive editions of *A World bibliography of bibliographies* why I decided to arrange it alphabetically by specific subjects. Since that decision was taken, and after prolonged experience of the book in use, I have had no reason to regret it, nor among the many letters I have received from librarians has there been a single one complaining of the alphabetical form of the *World bibliography*.

The *World bibliography of bibliographies* covers all subjects and all languages, and is intended to serve reference and research purposes of the most specific and specialised kind. Yet contained in it are broad and detailed surveys which, if relevant entries throughout the volumes are added to them, can serve also the widest reference inquiries, and be useful to those who seek primary signposts to information in varied fields of inquiry.

Therefore I can only thank Rowman and Littlefield for having gathered together all the titles in some of the major fields found throughout the 6664 columns of the fourth edition (1965-1966) of *A World bibliography of bibliographies*.

Preface

These fields are:
1. Bibliography
2. Printing
3. Periodical Publications
4. Academic Writings
5. Art and Architecture
6. Music and Drama
7. Education
8. Agriculture
9. Medicine
10. Law
11. English and American Literature
12. Technology
13. Physical Sciences
14. Biological Sciences
15. Family History
16. Commerce, Manufactures, Labour
17. History
18. Geography

Of course these categories by no means exhaust the 117,000 separately collated volumes set out in the *World bibliography*, and the above titles will be added to if librarians wish for it.

Th. B.

Notes on the Arrangement

An Alternative to critical annotation

Consider what it is we look for in a normal bibliography of a special subject. Reflection will show, I think, that we look, above all, for completeness, just as we do in a bibliography of bibliographies. We desire completeness even more than accuracy (painfully uncongenial though it is for me to make such a statement); for in most cases a bibliography is intended to give us particulars of publications to which we wish to refer; thus we can always judge for ourselves (waiving gross errors) whether the bibliographer has correctly described these publications. On the other hand, anything that is omitted is lost until rediscovered.

The question is, therefore, whether it is possible to give some indication of the degree of completeness of a bibliography without indulging in the annotation which is impossible in a work of the present scope and scale. It seemed to me that this could be achieved, to a considerable extent, by recording the approximate number of entries set out in it. This method is, of course, a rough-and-ready one, but experience shows that it is remarkably effective: and I hope that its novelty will not tell against it.

The recording of the number of works set out in a bibliography has another advantage in the case of serial publications: it displays in statistical form the development of the subject from year to year—often in a highly significant manner.

This procedure, then, is that which I have adopted, the number of items in each bibliography being shown in square brackets at the end of the entry. This, I may add, is by no means an easy or mechanical task, as can be judged from the fact that this process, on the average, just about doubles the time taken in entering each bibliography.

Supplementary information in footnotes

I have said that this method of indicating the number of entries is intended to replace critical treatment; but it is not possible to exclude annotation altogether, for a certain minimum of added information is indispensable. Consequently many of my entries will be seen to have footnotes, in which the following types of information are recorded: a few words of explanation where the title is inaccurate, misleading, obscure, or in-

Notes

sufficiently informative; a statement to that effect where a work is in progress, where intermediate volumes in a series have not been published, or where no more have been published; an attempt to clarify complicated series; a note that a book was privately printed or in a limited number of copies, where this does not exceed 500, or in some abnormal manner, as on one side of the leaf, on coloured paper, or in a reproduction of hand-writing, or with erratic pagination; when I have come across copies containing manuscript or other added matter, I have recorded the fact; substantial corrections and additions to bibliographies are sometimes published in periodicals, and I have noted a good many of these—but without aiming at anything even remotely approaching completeness, the attainment of which would be impossible. Various minor types of information are also occasionally to be found in the footnotes.

Owing to the great increase in the number of bibliographies reproduced directly from type-written copy, such publications are designated by an asterisk at the end of the entry; this device saves a good deal of space.

Place of publication

The place of publication is not shown when it is London in the case of an english book and

Notes

Paris in the case of a french one. In the case of a series or sequence of entries, however, the absence of a place of publication means that it is the same as the place last shown in the series. The same applies to the names of editors and compilers. The place of publication is given as it appears on the titlepage, but prepositions are omitted even if violence is done to grammatical construction.

The Order of entries

Under each heading the order of the entries is chronological by date of publication; in the case of works in successive volumes or editions the chronological order applies to the first volume or edition. In suitable long headings an additional chronological order by period covered has been created; see, for instance, France: History, or Drama: Great Britain.

Method of collating

An effort has been made, so far as space allows, to give detailed and accurate information of the kind more usually found in small bibliographies. For instance, I have paid special attention to the collation of bibliographies in several (or even numerous) parts or volumes. It is, in fact, difficult to understand why it is usually considered necessary to give collations of works in a single volume,

Notes

where difficulties seldom occur (from the point of view of systematic bibliography), but not of a work in several volumes, where confusion much more frequently arises. An occasional gap in the collations of such publications will be noticed. This is because, naturally enough, I have not been able in every case to see perfect sets; and I have thought it better to leave a very small number of such blanks rather than to hold up the bibliography indefinitely.

Serial publications

Where successive issues of a serial publication are set out, the year or period shown is usually that covered by the relevant issue; in such cases no date of publication is given unless publication was abnormal or erratic in relation to the period covered.

Bibliographies in more than one edition

Where a bibliography has gone into more than one edition I have tried (though I have not always been able) to record at least the first and latest editions. Intermediate editions have also been recorded wherever it seemed useful to do so, that is, for bibliographies first published before 1800, and for those of special interest or importance; but in general intermediate editions, though examined, have not been recorded.

Notes

Transcription of titles

Titles have been set out in the shortest possible form consistent with intelligibility and an adequate indication of the scope of the bibliography; omissions have of course been indicated. The author's name, generally speaking, is given as it appears on the titlepage, amplified and interpreted within square brackets where necessary.

Anonymous bibliographies

Far too large a proportion of bibliographical work is published anonymously. This is due, in part, to the all too common practice of library committees and similar bodies of suppressing altogether or of hiding in prefaces the names of those who have compiled bibliographies and catalogues for them. I have spent a good deal of time in excavating such and other evidences of authorship, and the result may be seen in the large number of titles preceded by names enclosed within square brackets.

Th. B.

Anatomy.

JAMES DOUGLAS, Bibliographiæ anatomicæ specimen: sive catalogus omnium penè auctorum qui ab Hippocrate ad Harveum rem anatomicam . . . scriptis illustrârunt. Londini 1715. pp.[xx].226.[x]. [1500.]

—— Editio secunda. [Edited by Adrianus Peleryn Chrouet]. Lugduni Batavorum 1734. pp.[xxiv].263.[xii]. [2000.]

ANDREAS OTTOMAR GOELICKE, Historia anatomiae nova aeqve ac antiqva, seu conspectus plerorumque, si non omnium, tam veterum quam recentiorum scriptorum, qui a primis artis medicae originibus suque ad praesentia nostra tempora anatomiam operibus suis illustrarunt. Halae Magdeburgicae 1713. pp.[x].224.[xxx.] [500.]

—— [another edition]. Introductio in historiam litterariam anatomes [&c.]. Francofurti ad Viadrum 1738. pp.[iii].12.540.[xxx]. [1000.]

[BARON] ALBRECHT VON HALLER, Bibliotheca anatomica, quae scripta ad anatomen et physiologiam facientia a rerum initiis recensentur. Tomus I. Ad annum MDCC. [II. Ab anno MDCCI. ad

I

MDCCLXXVI.] Tiguri 1774–1777. pp.viii.816+
[iii].870. [15,000.]

also issued with a London imprint.

—— Christophori Theophili de Mvrr Adno-
tationes ad Bibliothecas Hallerianas botanicam,
anatomicam, chirvrgicam et medicinae practicae.
Erlangae 1805. pp.[iii].67. [300.]

CHRISTIAN LUDWIG SCHWEICKHARD, Tentamen
catalogi rationalis dissertationum ad anatomiam
et physiologiam spectantium ab anno MDXXXIX.
ad nostra usque tempora. Tubingae 1798. pp.vi.
444. [3328.]

VOLSTÄNDIGES verzeichniss der literatur der
allgemeinen anatomie der menschen und der
haussäugethiere. Bern &c. 1840. pp.[ii].35. [790.]

[ÉMILE BLANCHARD], Notice sur les travaux
d'anatomie . . . de m. Émile Blanchard. 1850.
pp.[ii].36. [54.]

[JEAN LOUIS ARMAND DE QUATREFAGES DE BRÉAU],
Notice sur les travaux zoologiques et anatomiques
de m. A de Quatrefages. [1850]. pp.56. [78.]
—— [another edition]. 1852. pp.56. [84.]

[CLAUDE BERNARD], Notice sur les travaux
d'anatomie . . . de m. Claude Bernard [1850.]
pp.38. [31.]

— — [another edition]. Notice sur les travaux de m. Claude Bernard. 1854. pp.[ii].42. [45.]

[JOHANN] LUDWIG CHOULANT, Geschichte und bibliographie der anatomischen abbildung nach ihrer beziehung auf anatomische wissenschaft und bildende kunst. Leipzig 1852. pp.[xiii].xix.204. [400.]

— — History and bibliography of anatomic illustration. . . . Translated and edited by Mortimer Frank. Chicago 1920. pp.xxvii.435. [500.]
a revised facsimile of the translation appeared New York 1962.

J. VICTOR CARUS, Jahresbericht über die im gebiete der zootomie erschienenen arbeiten. . . . I. Bericht über die Jahre 1849–1852. Leipzig 1856. pp.viii.228. [1000.]
no more published.

PAUL BERT, Revue des travaux d'anatomie et de physiologie publiés en France pendant l'année 1864. Congrès des sociétés savantes: Caen 1865. pp.61. [150.]

ADOLPH BÜCHTING, Bibliotheca anatomica et physiologica, oder verzeichniss aller . . . in den letzten 20 jahren . . . im deutschen buchhandel erschienenen bücher und zeitschriften. Nordhausen 1868. pp.85. [850.]

BERICHT über die fortschritte der anatomie und physiologie. Herausgegeben von J. Henle, W[ilhelm] Keferstein und G[eorg] Meissner. Leipzig &c.

    1868. pp.iv.512. [1000.]
    1869. pp.vi.538. [1000.]
    1870. pp.iv.324. [750.]
    1871. pp.v.358. [750.]
no more published.

JAHRESBERICHTE über die fortschritte der anatomie und physiologie. Leipzig.

    i. 1872. Edited by Fr[anz Adolf] Hofmann and G. Schwalbe. 1873. pp.ix.612. [1250.]
    ii. 1873. 1875. pp.vii.574. [1500.]
    iii. 1874. 1875. pp.vi.446.282. [1750.]
    iv. 1875. 1876. pp.vi.450.276. [1750.]
    v. 1876. 1878. pp.vi.587.316. [2250.]
    vi. 1877. 1878. pp.v.458+iv.236+iv.342. [2500.]
    vii. 1878. 1879. pp.vi.470+iv.247+iv.432. [2500.]
    viii. 1879. 1880. pp.vi.458+iv.410. [2500.]
    ix. 1880. 1881. pp.iv.456+iv.522. [2750.]
    x. 1881. 1882. pp.iv.484+iv.458. [2750.]
    — General-register . . . i-x band. [1886]. pp. 164.

xi. 1882. 1883. pp.iv.400+442. [3000.]

xii. 1883. 1884–1885. pp.iv.494+iv.490. [3500.]

xiii. 1884. 1886. pp.iv.562+iv.482. [4000.]

xiv. 1885. 1886–1887. pp.iv.630+iv.476. [4500.]

xv. 1886. Edited by L[udimar] Hermann and G. Schwalbe. 1887–1888. pp.vi.650+ iv.432. [5000.]

xvi. 1887. 1888. pp.iv.824+iv.424. [5000.]

xvii. 1888. 1889. pp.iv.738+iv.480. [5000.]

xviii. 1889. 1890. pp.iv.694+iv.512. [4500.]

xix. 1890. 1891. pp.iv.812+iv.550. [5500.]

xx. 1891. 1892–1893. pp.iv.780+iv.732. [4500.]

no more published; but the physiological part was continued by the Jahresbericht über die fortschritte der physiologie, *which is entered under Physiology, below; in the present work an index to vols.xi–xx forms part of vol.xx.*

LIST of books in the National art library . . . on anatomy, human and comparative. South Kensington museum [*afterwards:* Victoria and Albert museum]: 1886. pp.24. [300.]

ANATOMISCHE schriften des professors d^r Paul Albrecht. Hamburg 1886. Broadside. [73.]
— [second edition]. 1886. pp.viii. [91.]

Anatomy

ANATOMISCHE hefte. . . . Zweite abteilung.
Ergebnisse der anatomie und entwicklungs-
geschichte. . . . Herausgegeben von Fr[iedrich
Siegmund] Merkel und R[obert] Bonnet. Wiesbaden.

    i. 1891. pp.xix.778. [2500.]
    ii. 1892. pp.xi.670. [3000.]
    iii. 1893. pp.xi.634. [2500.]
    iv. 1894. pp.xi.660. [3000.]
    v. 1895. pp.x.732. [3500.]
    vi. 1896. pp.viii.812. [3500.]
    vii. 1897. pp.xi.990. [3500.]
    viii. 1898. pp.xi.1166. [4500.]
    ix. 1899. pp.x.886. [4000.]
    x. 1900. pp.xv.1134. [4000.]
    xi. 1901. pp.xii.1188. [4000.]
    xii. 1902. pp.xi.932. [5000.]
    xiii. 1903. pp.ix.748. [3000.]
    xiv. 1904. pp.xi.1310. [5000.]
    xv. 1905. pp.ix.1014. [4000.]
    xvi. 1906. pp.xii.944. [3500.]
    xvii. 1907. pp.x.768. [3500.]
    xviii. 1908. pp.viii.906. [3500.]
    xix. 1909. pp.xi.526+x.527–1123. [5000.]
    xx. 1911. pp.vi.469+viii.471–1251. [5500.]
    xxi. 1913. pp.vii.372. [2000.]
    xxii. 1914. pp.xxiii.635. [3000.]
    [*continued as:*]

Zeitschrift für die gesamte anatomie. . . . Dritte
abteilung. Ergebnisse [&c.] . . . . Unter mitwirkung
von F[ranz] Heiderich . . . herausgegeben von
Erich Kallius. Münster &c.

xxiii. 1921. pp.vii.925. [4000.]
xxiv. 1923. pp.vii.752. [2500.]
xxv. 1924. pp.x.833. [3000.]
xxvi. 1925. pp.viii.520. [2000.]
xxvii. 1927. pp.x.1104. [4500.]

in the new series the dates are those of publication.

BIBLIOGRAPHIE anatomique. Revue des travaux
en langue française. Sous la direction de m.
A[dolphe] Nicolas.

i. 1893. pp.[iii].203. [705.]
ii. 1894. pp.[iii].236. [804.]
iii. 1895. pp.[iii].286. [714.]
iv. 1896. pp.[iii].286. [768.]
v. 1897. pp.[iii].292. [657.]
vi. 1898. pp.[iii].328. [668.]
vii. 1899. pp.[iii].314. [604.]
viii. 1900. pp.[iii].316. [738.]
ix. 1901. pp.[iii].336. [532.]
x. 1902. pp.[iii].314. [555.]
xi. 1902. pp.[iii].320. [266.]
xii. 1903. pp.[iii].320. [737.]
xiii. 1904. pp.[iii].316. [770.]
xiv. 1905. pp.[iii].322. [537.]

xv. 1906. pp.[iii].318. [454.]
xvi. 1907. pp.[iii].324. [777.]
xvii. 1908. pp.[iii].306. [363.]
xviii. 1909. pp.[iii].308. [416.]
xix. 1909[-1910]. pp.[iii].321. [386.]
xx. 1910. pp.[iii].348. [190.]
xxi. [1910-]1911. pp.[iii].322. [569.]
xxii. 1912. pp.[iii].314. [560.]
xxiii. 1913. pp.[iii].314. [673.]
xxiv. 1914. pp.[iii].318. [386.]
xxv. 1914-1918. pp.129. [427.]
— Tables. Dressés par A. Hovelacque.
Nancy 1928. pp.64.
no more published.

ANATOMISCHER anzeiger. Litteratur. Jena.
[1895]. Von E[rnst Carl Ferdinand] Roth.
pp.ccxvi. [3000.]
[*continued as:*]
Anatomischer anzeiger. Bibliographia anatomica.
[1896]. Edit Concilium bibliographicum.
pp.104. [1500.]
[1897]. pp.132. [2000.]
[*continued as:*]
Anatomischer anzeiger. Litteratur.
1898. Von Otto Hamann, pp.cciv. [3500.]
1899. pp.112+112. [4000.]

1900. pp.112+112. [4000.]

1901. pp.112+104. [4000.]

1902. pp.80+104. [3000.]

1903. pp.112+96. [3500.]

1904. pp.128+96. [3500.]

1905. pp.96+112. [3500.]

1906. pp.96+112. [3500.]

1907. pp.96+112. [3500.]

1908. pp.96+80. [3000.]

1909. pp.112+64. [3000.]

1910. pp.80+64. [2500.]

1911. pp.64+80. [2500.]

1911–1912. pp.80+64. [2500.]

1912–1913. pp.48+64. [2000.]

1913–1914. pp.64+32. [1500.]

1914–1915. pp.64+64. [2000.]

1915–1917. pp.48+48. [1500.]

1917–1919. pp.32+48. [1250.]

1920–1921. pp.32+80. [2000.]

1921–1922. pp.64+64. [2000.]

1923–1924. pp.80+32+32. [2500.]

1925. [*not published*].

1926. Von Fritz Prinzhorn. pp.96+128. [3500.]

1927. pp.208+144. [5000.]

earlier issues form part of the main work.

JAHRESBERICHT über die leistungen und fortschritte in der anatomie und physiologie. Herausgegeben von Rudolf Virchow. Berlin.

> 1897. Unter special-redaction von E. Gurlt
> und C. Posner. pp.[iii].234. [3000.]
>
> 1898. Unter redaction von C. Posner.
> pp.[iii].246. [3250.]

ROBERT VON TÖPLY, Studien zur geschichte der
anatomie im mittelalter. Leipzig &c. 1898.
pp.vii.121. [500.]

HUMAN anatomy [4th–14th issues: Anatomy].
International catalogue of scientific literature
(section O): Royal society.

> i. 1903. pp.xiv.212. [1309.]
> ii. 1904. pp.viii.235. [1424.]
> iii. 1905. pp.viii.228. [1529.]
> iv. 1906. pp.viii.212. [1207.]
> v. 1907. pp.viii.400. [2543.]
> vi. 1908. pp.viii.318. [2070.]
> vii. 1909. pp.viii.308. [2652.]
> viii. 1911. pp.viii.241. [1968.]
> ix. 1912. pp.viii.276. [2614.]
> x. 1913. pp.viii.244. [2618.]
> xi. 1915. pp.viii.276. [2810.]
> xii. 1916. pp.viii.268. [2659.]
> xiii. 1917. pp.viii.157. [1399.]

xiv. 1919. pp.viii.171. [1389.]
no more published.

[EMANUEL LEFAS], Titres et travaux d'anatomie pathologique par . . . E. Lefas. 1910. pp.101. [92.]

CATALOGUS van de boeken en tijdschriften geplaatst in de laboratoria voor ziektekundige ontleedkunde, gezondheidsleer en ontleedkunde. Universiteit: Bibliotheek: Amsterdam 1918. pp.[iii].52. [850.]

BIBLIOGRAPHIC service [*afterwards:* The Wistar institute bibliographic service]. Wistar institute of anatomy and biology: Philadelphia, Pa.

i. 1917–1919. pp.227. [519.]
ii. 1920–1922. pp.221. [527.]
iii. 1923–1925. pp.x.328. [854.]
iv. 1926–1927. Edited by Helen Dean King. pp.x.316. [744.]
v. 1928–1929. pp.x.384. [1027.]
vi. 1930–1931. pp.x.353. [867.]
vii. 1932–1933. pp.x.377. [885.]
viii. 1934–35. pp.x.441. [957.]

no more published; this work comprises abstracts of papers published in the Institute's journals.

EDGAR GOLDSCHMID, Entwicklung und bibliographie der pathologisch-anatomischen abbildung. Leipzig 1925. pp.[v].503. [500.]

Anatomy

PHILIPP STÖHR, Anatomy, histology and embryology. Office of military government for Germany: Field information agencies technical: Fiat review of german science, 1939–1946: Wiesbaden 1947. pp.[xi].148. [75.]
the text is in german.

HANS HELLER, Bibliographie der basler anatomischen arbeiten von 1537–1900. Acta anatomica: Supplementa (no.11): Basel &c. 1949. pp.54. [484.]

P[ETR] O[SIPOVICH] ISAEV, Библиография отечественной литературы по анатомии человека. Москва 1962. pp.452. [12,000.]

K[ENNETH] F[ITZPATRICK] RUSSELL, British anatomy 1525–1800. A bibliography. [Parkville, Victoria 1963]. pp.iii–xvii.254. [901.]

Anthropology and ethnology.

5. 'Race', 50.
6. Teaching, 51.
7. Miscellaneous, 52.

1. *Periodicals and congresses*

LISTE des publications périodiques reçues régulièrement. Musée de l'homme: Bibliothèque: 1952. ff.68. [500.]*
— [another edition]. 1956. ff.[i].103. [800.]*

PUBLICAÇÕES periódicas estrangeiras inventariadas nas bibliotecas portuguesas. Etnologia, ciências naturais, agro-pecuária. Instituto de alta cultura: Centro documentação científica: Lisboa 1953. pp.739. [8000.]

ANTHROPOLOGICAL journals and monograph series . . . in the libraries of Great Britain. Royal anthropological institute: [1953]. pp.[ii].24. [300.]*
— [another edition]. Survey of anthropological journals and monograph series in libraries in the United Kingdom. [1957]. pp.17. [600.]

JUAN COMAS [CAMPS], Historia y bibliografía de los congresos internacionales de ciencias antropológicas: 1865–1954. Universidad nacional antónoma de México: Instituto de historia: Publicaciones (1st ser., no.37): México 1956. pp.470. [4000.]

HANS JÜRGEN ASCHENBORN, Anthropological and archaeological journals in south african libraries. State library: Bibliographies (no.2): Pretoria 1961. ff.[18]. [150.]*

2. General

VERZEICHNISS der anthropologischen literatur. [Archiv für anthropologie: Supplement]. Braunschweig.
> [1872]. pp.47. [1000.]
> [1873]. pp.61. [1250.]
> [1874]. pp.65. [1500.]
> [1875]. pp.69. [1750.]
> [1876]. pp.83. [2250.]
> [1878]. pp.97. [2500.]
> [1880]. pp.141. [4000.]
> [1880]. pp.153. [4500.]
> [1881]. pp.143. [4000.]
> [1883]. pp.161. [5000.]
> [1885]. pp.156. [4500.]
> [1886]. pp.135. [4000.]
> [1888]. pp.194. [5500.]
> [1889]. pp.152. [4500.]

[1891]. pp.164. [4500.]
[1891]. pp.161. [4500.]
[1893]. pp.141. [4000.]
[1894]. pp.130. [3500.]
[1895]. pp.160. [5000.]
[1897]. pp.182. [5500.]
[1898]. pp.244. [7000.]
[1899–1900]. pp.199. [5000.]
[1902]. pp.210. [5000.]
[1904]. pp.169. [4500.]
[1904]. pp.138. [3500.]

the dates are those of publication; forms a supplement to vols.v–n.s.i of the Archiv; *four earlier issues formed part of it; no more published.*

v. [WILHELM DAVID] KONER, Литература антропологіи, этнологіи и науки о древностяхъ за 1876 годъ. С.-Петербургъ 1878. pp.32. [850.]

PAOLO RICCARDI, Saggio di un catalogo bibliografico antropologico italiano. Modena 1883. pp.[v].176. [750.]

GEORGE W. BLOXAM, Index to the publications of the Anthropological institute of Great Britain and Ireland [1843–1891], including the Journal and Transactions of the Ethnological society . . . ,

17

the Journal and Memoirs of the Anthropological society . . . ; the Anthropological review; and the Journal of the Anthropological institute. 1893. pp.viii.302. [2500.]

CATALOG der handbibliothek des Königlichen zoologischen und anthropologisch-ethnographischen museums in Dresden. Berlin 1898. pp.xxv. 287. [anthropology: 2750.]

ČENĚK ZÍBRT, Literatura kulturně-historická a ethnografická, 1897–1898. Praze 1898. pp.92.xxxi. [894.]

PHYSICAL anthropology [3rd–14th issues: Anthropology]. International catalogue of scientific literature (section P): Royal society.

    i. 1903. pp.xiv.224. [1545.]
    ii.1904. pp.viii.299. [1861.]
    iii. 1905. pp.viii.324. [2095.]
    iv. 1906. pp.viii.411. [2691.]
    v. 1907. pp.viii.423. [2614.]
    vi. 1908. pp.xix.444. [3260.]
    vii. 1910. pp.viii.405. [3492.]
    viii. 1911. pp.viii.508. [4395.]
    ix. 1912. pp.viii.427. [4076.]
    x. 1913. pp.viii.472. [4268.]
    xi. 1914. pp.viii.419. [4118.]
    xii. 1915. pp.viii.489. [4943.]

xiii. 1917. pp.viii.450. [4434.]
xiv. 1919. pp.viii.380. [3244.]
no more published.

NORTHCOTE W[HITRIDGE] THOMAS, Bibliography of anthropology and folk-lore ... containing works published within the British Empire.

1906. 1907. pp.lxxii. [779.]
1907. 1908. pp.lxxiv. [874.]
no more published.

JUUL DIESERUD, The scope and content of the science of anthropology. Chicago 1908. pp.200. [500.]

J. C. E. SCHMELTZ, Catalogus van 's Rijks ethnographisch museum. Deel III. Catalogus der bibliotheek. Leiden 1909. pp.v–xxi.279. [5000.]

S[EBALD]R[UDOLF]STEINMETZ, Essai d'une bibliographie systématique de l'ethnologie jusqu'à l'année 1911. Institut de sociologie: Monographies bibliographiques (no.1): Bruxelles &c. [1912]. pp.iv.196. [2000.]

CURRENT anthropological literature. American anthropological association *and* American folklore society: Lancaster, Pa.

i. 1912. pp.[ii].351. [1500.]
ii. 1913. pp.[ii].256. [1000.]
no more published.

THE ATHENÆUM subject index to periodicals, 1915. . . . History, geography, anthropology, and folk-lore. 1916. pp.32. [1000.]

— 1916. Historical, political, and economic sciences, including . . . anthropology. 1917. pp.120. [4000.]

no more sections devoted to anthropology were published.

HANS PLISCHKE, Anthropologie, vorgeschichte, völkerkunde. Jahresberichte des Literarischen zentralblattes über die wichtigsten wissenschaftlichen neuerscheinungen des gesamten deutschen sprachgebietes (1924, vol.xvii): Leipzig 1925. pp.69. [600.]

ETHNOLOGISCHER anzeiger. [Jahresbibliographie und bericht über die völkerkundliche literatur]. Stuttgart.

    i. [1924–1925]. Herausgegeben von M. Heydrich. 1928. pp.vi.504. [5016.]

    ii. [1926–1927]. 1929–1932. pp.[iii].482.366. [8603.]

    iii. [1928–1935]. 1932–1935. pp.vii.485.348. [9179.]

    iv. [1936–1940]. pp.352. [7019.]

incomplete; no more published.

FALK [ALFRED] RUTTKE, Schrifttum und aufklärungsstoff zur volkspflege. Rassenkunde &c. Reichsausschuss für volksgesundheitsdienst. Schriftenreihe (no.8): Berlin 1935. pp.24. [250.]

WERNER and ANNE MARIE QUENSTEDT, Hominidæ fossiles. Fossilium catalogus (1. 74): 's Gravenhage 1936. pp.456. [12,500.]

ELIZABETH G[REENEBAUM] HERZOG, General index, American anthropologist, Current anthropological literature, and Memoirs of the American anthropological association, 1929-1938. American anthropologist (vol.xlii, no.4, part 3): [Menasha, Wis.] 1940. pp.[v].122. [1700.]

A BRIEF list of recent books on anthropology and ethnology. Library of Congress: Washington 1940. ff.2. [20.]*

JUAN COMAS [CAMPS], Recopilación bibliográfica de antropología prehistórica. Homo neanderthalensis. México 1941. pp.[12]. [180.]

ANTHROPOLOGY. A selected list of books compiled by the Royal anthropological institute. National book council: Book list (no.185): 1943. pp.[4]. [175.]

MAMIE TANQUIST MILLER, An author, title, and subject check list of Smithsonian institution publications relating to anthropology. University of

New Mexico: Bulletin (no.405 = Bibliographical series, vol.i, no.2): Albuquerque 1946. pp.218. [1250.]

HENRY FIELD, List of documents microfilmed, 1941–1948. Field museum of natural history: Washington 1949. pp.13. [37.]

REFERENCES on living and working conditions of indigenous populations. International labour office: Bibliographical reference list (no.3). Geneva 1950. pp.17. [300.]*
— Revised. 1950. pp.31. [500.]*
— — Supplement. 1950. pp.4. [50.]*

LISTE des acquisitions de la bibliothèque. Musée de l'homme.*
    1951. ff.29. [450.]
    1952. ff.36. [600.]
    1953–1955. 1956. ff.72. [1400.]
    1957. ff.[i].40.[600.]
    1959. ff.48. [700.]
in progress.

A. L. KROEBER, *ed.* Anthropology today. An encyclopedic inventory. Chicago 1953. pp.xv.966. [2500.]
— — Current anthropology. A supplement. 1956. pp.xii.377. [2000.]

Anthropology

LUDVÍK KUNZ, Česká ethnografie a folklorista v letech 1945–1952. Československá akademie věd: Sekce filosofie a historie: Praha 1954. pp.384. [2721.]

INTERNATIONAL bibliography of socio–cultural anthropology. [Paris].*
  i. [1954.] [By L. Bernot]. ff.[x].385.[xix]. [3327.]*

[continued as:]
International bibliography of social and cultural anthropology. Prepared by the International committee for social sciences documentation [&c.]. Unesco: [Paris].

  i. 1955. Editors: Georges Balandier [*and others*]. pp.259. [3563.]
  ii. 1956. Editor: Pierre Clément. pp.391. [4512.]
  iii. 1957. Editor: J. F. M. Middleton. pp.410. [5146.]
  iv. 1958. pp.341. [4896.]
  v. 1959. pp.443. [6198.]
  vi. 1960. 1962.
  vii. 1961. 1963. pp.xxxviii.254. [3646.]
  viii. 1962. 1964. pp.xxxii.175. [2450.]
in progress; also issued in a french edition.

Anthropology

BIBLIOGRAPHIE sur la méthode statistique appliquée à la médecine et aux sciences anthropologiques. Liste d'ouvrages et d'articles en langue
française. Centre d'études statistiques de la chaire
d'hygiène et de médecine préventive and Institut
national de la statistique et des études économiques: 1955. pp.34. [200.]*

CUMULATIVE index to current literature on
anthropology & allied subjects. Department of
anthropology: Calcutta.*
 i. 1955. pp.[viii].171. [1500.]
 ii. 1956. pp.[iv].144. [1250.]

BIENNAL review of anthropology. Stanford, Cal.
 1959. Edited by Bernard J[oseph] Siegel.
 pp.ix.273. [1450.]
 1961. pp.[ix].338. [1700.]
 1963. pp.[ix].315. [1500.]
in progress.

ADAM WRZOSEK, Bibliografia antropologii polskiej do roku 1955 włącznie. Polska akademia nauk:
Zakład antropologii: Materiały i prace antropolo-
giczne (no.41): Wrocław 1959–1960. pp.468+
203. [4854.]

ANNA WIRPSZA, Zestawienie nabytków Polskiego towarzystwa Ludoznawczego za okres 1956–1959. Polskie towarzystwo ludoznawcze: Archiwum etnograficzne (no.23): Wrocław 1960 [1961]. pp.145. [1933.]*

новая иностранная литература по истории, археологии и этнографии. Академия наук СССР: Фундаментальная библиотека общественных наук: Москва 1960 &c.
in progress.

новая советская литература по истории, археологии и этнографии. Академия наук СССР. Фундаментальная библиотека общественных наук: Москва 1960 &c.
in progress.

ANDREW KYLE and GILBERT E[UGENE] DONAHUE, Guide to the sources and publishing agencies of literature in anthropology. Wayne state university: Monteith college: Working bibliography series (no.2): Detroit 1961. ff.iii.17. [150.]*

JANE CLAPP, Museum publications. Part 1. Anthropology, archaeology and art. New York 1962. pp.434. [4416.]*

AUTHOR and subject catalogues of the library of

the Peabody museum of archaeology and ethnology, Harvard university. Author catalogue.
Boston 1963.

    i. A–Arj. pp.[iv].556. [11,676.]

    ii. Ark–Bens. pp.[ii].560. [11,760.]

    iii. Bent–Brer. pp.[ii].548. [11,508.]

    iv. Bres–Cart. pp.[ii].546. [11,466.]

    v. Caru–Con. pp.[ii].548. [11,608.]

    vi. Cons–Deus. pp.[ii].550. [11,500.]

    vii. Deut–Eust. pp.[ii].550. [11,500.]

    viii. Eva–Gaba. pp.[ii].550. [11,500.]

    ix. Gabb–Grev. pp.[ii].548. [11,608.]

    x. Grew–hedg. pp.[ii].548. [11,608.]

    xi. Hedi–Hum. pp.[ii].550. [11,550.]

    xii. Hun–Kall. pp.[ii].550. [11,550.]

    xiii. Kalm–Lada. pp.[ii].550. [11,550.]

    xiv. Ladd–Lonb. pp.[ii].546. [11,466.]

    xv. Lond–Mar. pp.[ii].550. [11,550.]

    xvi. Mas–Mont. pp.[ii].568. [11,928.]

    xvii. Monu–Noi. pp.[ii].554. [11,634.]

    xviii. Nok–Pem. pp.[ii].560. [11,760.]

    xix. Pen–Ramc. pp.[ii].582. [12,222.]

    xx. Rame–Rudn. pp.[ii].580. [12,180.]

    xxi. Rudo–Selk. pp.[ii].577. [12,117.]

    xxii. Sell–Spon. pp.[ii].564. [11,844.]

    xxiii. Spoo–Thomond. pp.[ii].552. [11,592.]

    xxiv. Thompson–Vavp. pp.[ii].547. [11,487.]

xxv. Vayda–Wids. pp.[ii].542. [11,382.]
xxvi. Wie–Z. pp.[ii].415. [8715.]
— Subject catalogue.

i. A–America (archaeology). pp.[iv].612. [12,852.]

ii. America (atlases)–Archaeology (pic). pp.[ii].617. [12,957.]

iii. Archaeology (pig)–Australia (archaeology: jade). pp.[ii].603. [12,663.]

iv. Australia (archaeology: knives)–Burma ethnology). pp.[ii].567. [11,907.]

v. Burma–F–Central America (L). pp.[ii].577. [12,117.]

vi. Central America (M)–Dobu Island. pp.[ii].576. [12,096.]

vii. Dodecanese–Ethnology. pp.[ii].573. [12,033.]

viii. Ethnology (A)–France (archaeology: Fa). pp.[ii].566. [11,886.]

ix. France (archaeology: Fi)–Germany (archaeology: burial). pp.[ii].554. [11,634.]

x. Germany (archaeology: camps)–Hungary (archaeology: Me). pp.[ii].532. [11,172.]

xi. Hungary (archaeology: Mi)–Ire. pp.[ii].539. [11,319.]

xii. Iringa province–Kiowan. pp.[ii].537. [11,277.]

xiii. Kiplingu–Mexico (archaeology). pp.[ii]. 555. [11,655.]

xiv. Mexico (archaeology)–Musical instruments. pp.[ii].550. [11,550.]

xv. Muskhogan–Niénégué. pp.[ii].548. [11,508.]

xvi. Niger Coast Protectorate–North America (southwest linguistics). pp.[ii].581. [12,201.]

xvii. North America (southwest maps)–Phallicism. pp.[ii].571. [11,991.]

xviii. Philae–Religion (symbolism). pp.[ii]. 575. [12,075.]

xix. Religion (taboo)–Siouan–Dakota (traditions). pp.[ii].584. [12,264.]

xx. Siouan–Hidatsa–Somatology (chemistry). pp.[ii].615. [12,915.]

xxi. Somatology (child study)–Somatology (fossil man–Weimar). pp.[ii].540. [11,340.]

xxii. Somatology (frontal bone)–Somatology (placenta). pp.[ii].540. [11,340.]

xxiii. Somatology (platyrrhina)–South Dakota (zoology). pp.[ii].440. [9240.]

xxiv. Southampton Island–Techialoyan. pp. [ii].476. [9996.]

xxv. Technology (Antoine)–Technology (magnifying). pp.[ii].512. [10,752.]

xxvi. Technology (map making)–Tibet (traditions). pp.[ii].537. [11,277.]

xxvii. Tibetans (somatology)–Z. pp.[ii].534. [11,214.]

reproduces the catalogues cards, which are represented by the figures in square brackets.

DAVID G[OODMAN] MANDELBAUM [*and others*], Resources for the teaching of anthropology. Berkeley &c. 1963. pp.[v].316. [1714.]*

INDEX to current periodicals received in the library of the Royal anthropological institute.

i. 1963. pp.xxi.294. [5228.]
in progress.

3. *Countries &c.*

Africa

I[SAAC] SCHAPERA [*and others*], Select bibliography of south african native life and problems. Inter-university committee for african studies: 1941. pp.xii.250. [2206.]

— — Supplement. Modern status and conditions. [By] A. M. [Margaret Anne] Holden and A[nnette] Jacoby. University of Cape Town: School of librarianship: Bibliographical series: [Capetown] 1950. ff.iv.20.iv.32. [421.]*

A. M. Holden on the titlepage is merely a misprint.

— — Second supplement. Modern status and

condition. . . . By R. Giffen and J. Back. 1958.
pp.[75]. [425.]★

NATIVE peoples in theaters of war. [1]. North
Africa. Public library: Newark, N.J. [1942].
pp.[8]. [50.]
no more published.

D[OREEN] M. WOOKEY, A bibliography of physical anthropology in South Africa, 1936–1947.
University of Cape Town: School of librarianship: Bibliographical series: [Capetown] 1947.
pp.v.45. [191.]★

H. A. WIESCHHOFF, Anthropological bibliography of Negro Africa. American oriental society:
American oriental series (volume 23): New Haven
1948. pp.xi.461. [15,000.]★

[L. SANNER], Bibliographie ethnographique de
l'Afrique équatoriale française, 1914–1948. Haut
commissaire de la république en Afrique équatoriale française: 1949. pp.3–107. [549.]

WILFRID D[YSON] HAMBLY, Bibliography of
african anthropology, 1937–1949. Chicago natural
history museum: Fieldiana-anthropology (vol.
xxxvii, no.2): Chicago 1952. pp.155–292. [1500.]
forms a supplement to the bibliography in the same
author's Source book for african anthropology,
Chicago 1937.

Anthropology

ALFONSO COSTA, Bibliografia do etnólogo p^re Carlos Estermann. Instituto de Angola: Luanda 1961. pp.16. [62.]

Algeria

CAMILLE LACOSTE, Bibliographie ethnologique de la Grande Kabylie. Maison des sciences de l'homme: Recherches méditerranéennes: Bibliographies (no.i): La Haye &c. 1962. pp.104. [732.]

America

BOLETÍN bibliográfico de antropología americana. Instituto panamericano de geografía e historia: México.

    i. 1937. pp.3–287. [1500.]
    ii. 1938. pp.[iv].194. [1000.]
    iii. 1939. pp.[iv].258. [1500.]
    iv. 1940. pp.[iii].316. [2000.]
    v. 1941. pp.404. [2500.]
    vi. 1942. pp.[ii].268. [1500.]
    vii. 1943–1944. pp.[ii].288. [1500.]
    viii. 1945. pp.[ii].284. [1500.]
    ix. 1946. pp.[ii].352. [2000.]
    x. 1947. pp.[ii].360. [2000.]
    xi. 1948. pp.[ii].470. [2500.]
    xii. 1949. pp.[ii].307+312. [3000.]
    xiii. 1950. pp.[ii].317+301. [3000.]
    xiv. 1951. pp.[ii].306+289. [3000.]

xv–xvi. 1952–1953. pp.485+374. [4000.]
xvii. 1954. pp.336+287. {3000.]
xviii. 1955. pp.261+328. [3000.]
xix–xx. 1956–1957. pp.233+397. [3000.]
xxi–xxii. 1958–1959. 1961–1962. pp.239+
387. [3000.]
in progress.

GEORGE PETER MURDOCK, Ethnographic bibliography of north America. Yale university:
Department of anthropology: Yale anthropological studies (vol.i): New Haven 1941. pp.xvi.168.
[10,000.]*

— — 3rd edition. Behavior science bibliographies: 1960. pp.xxiii.393. [17,500.]*

MAMIE [RUTH] TANQUIST MILLER, An author,
title and subject check list of Smithsonian institution publications relating to anthropology.
University of New Mexico: Bulletin (no.405 =
Bibliographical series, vol.i, no.2): Albuquerque
1946. pp.218. [1250.]

A COLLECTION of books pertaining to the archæology, ethnology & anthropology of Mexico,
Guatemala and central America . . . presented
by Matilda Geddings Cray to the Museo
nacional de antropología de Guatemala. San Fran-

cisco 1948. pp.[iii].107. [500.]
200 copies printed.

JUAN COMAS [CAMPS], Bibliografía selectiva de las culturas indígenas de América. Instituto panamericano de geografía e historia: Publicación (no.166 = Comisión de historia: Bibliografías, no.1): México 1953. pp.xxviii.284. [2000.]

BERNARD J[OSEPH] SIEGEL, *ed.* Acculturation. Critical abstracts, north America. Stanford anthropological series (no.2): Stanford [1955]. pp. xiv.231. [94.]*

IGNACÍO BERNAL [GARCÍA PIMENTEL], Bibliografía de arqueología y etnografía, Mesoamérica y norte de México 1514–1960. Instituto nacional de antropología e historia: Memorias (no.7): México 1962. pp.iii–xvi.635. [13,990.]

FRANCIS A. RIDDEL, A bibliography of the Indians of central California. Department of parks and recreation: Division of beaches and parks: Ethnographic report (no.3): Sacramento 1962. ff.[i].ii.33. [400.]*

FRANCIS A. RIDDEL, A bibliography of the Indians of southern California. Department of parks and recreation: Division of beaches and parks:

Ethnographic report (no.4): Sacramento 1962. ff.[iii].18. [200.]★

FRANCIS A. RIDDEL, A bibliography of the Indians of northeastern California. Deparment of parks and recreation: Division of beaches and parks: Ethnographic report (no.5): Sacramento 1962. ff.[i].ii.10. [125.]★

TIMOTHY J. O'LEARY, Ethnographic bibliography of south America. Human relations area files: Behavior science bibliographies: New Haven 1963. pp.xxiv.389. [20,000.]★

Argentina

JULIÁN B[ERNARDO] CÁCERES FREYRE, Bibliografía anthropológica argentina 1940. Buenos Aires 1941. pp.8. [125.]

Asia

A[LEKSIEI] A[RSENIEVICH] IVANOVSKY, Періодическія изданія Сибири и средней Азіи 1789–1801 гг. Указатель статей по этнографіи. Библіографическія записки: Москва 1892. pp.[i].14. [480.]

E. A. VOZNESENSKAYA and A. B. PIOTROVSKY, Материалы для библиографии по антропологии и этнографии Казакстана и Среднеазиатских республик. Академия наук: Труды Комиссии по изучению племенного состава населения СССР и сопредельных стран (vol.xiv): Ленинград 1927. pp.[iii].ix. 249. [2780.]

ELIZABETH VON FÜRER-HAIMENDORF, An anthropological bibliography of south Asia, together with a directory of recent anthropological field work. École pratique des hautes études: Le monde d'outre-mer passé et présent (4th ser., vol.iii): 1958. pp.748. [5316.]

Australia and Tasmania

R. ETHERIDGE, Contributions to a catalogue of works, reports, and papers on the anthropological, ethnological and geological history of the australian and tasmanian aborigines. Part I. Department of mines: Geological survey of New South Wales: Memoirs (Palæontology, no.8): Sydney 1890–1895. pp.vii.31+vii.49+vii.40. [1500.] *incomplete; no more published.*

GEORGE F[RASER] BLACK, List of works relating to the aborigines of Australia and Tasmania. Public library: New York 1913. pp.56. [1400.]

ARNOLD R. PILLING, Aborigine culture history A survey of publications 1954–1957. Wayne state

university: Studies (no.11): Detroit 1962. pp.ix.
219. [600.]

JOHN GREENWAY, Bibliography of the australian
aborigines and the native peoples of Torres strait
to 1959. Sydney &c. 1963. pp.xv.420. [10,283.]

Azerbaijan

G. A. GULIEV, Библиография этнографии
Азербайджана (изданной на русском языке
до 1917 года). Баку 1962–   . pp.128+   .
in progress.

Brazil

HERBERT BALDUS, Bibliografia critica da etnologia brasileira. Commissão do IV centenário da
cidade de São Paulo: Serviço de commemorações
culturais: São Paulo 1954. pp.861. [1785.]

HERBERT BALDUS, Bibliografia comentada de
etnologia brasileira, 1943–1950. Série bibliográfica de estudos brasileiros (no.1): Rio de
Janeiro 1954. pp.142. [354.]

Bulgaria

IVAN KOEV and KHRISTO VAKAERLSKI, Zehn jahre
bulgarische ethnographie. . . . Bibliographie . . .
1943–1952. Wissenschaftliche informationen zur
volkskunde, altertumskunde und kulturgeogra-

phie aus dem östlichen Europa (no.5): München 1959. pp.20. [30.]

Carpathian basin

IRMA ALLODIATORIS, A Kárpát-medence antropológiai bibliográfiája. Akadémiai kiadó: Budapest 1958. pp.183. [2800.]

Chile

CARLOS E. PORTFR, Bibliografía chileña de anthropología y etnología. Buenos Aires 1910. pp.[ii].147–188. [200.]

Colombia

SERGIO ELÍAS ORTIZ [CORTÉS], Contribución a la bibliografía sobre ciencias etnológicas de Colombia. Escuela normal de occidente: Idearium: Supplemento (no.1): Pasto 1937. pp.66. [500.]

Congo

BIBLIOGRAPHIE ethnographique du Congo belge et des régions avoisinantes. Musée du Congo belge [Musée royal de l'Afrique centrale], Tervueren: Bureau de documentation ethnographique: Publicarions [1st ser.), Bibliographies: Bruxelles.

 i. 1925–1930. [By J. Maes and Olga Boone]. 1932. pp.xi.358. [3000.]

Anthropology

ii. I. 1931. 1933. pp.88. [750.]
ii. II. 1932. 1933. pp.112. [750.]
ii. III. 1933. 1935. pp.141. [600.]
ii. IV. 1934. 1935. pp.116. [600.]
ii. V. 1935. 1936. pp.125. [600.]
iii. I. 1936. 1937. pp.139. [750.]
iii. II. 1937. 1938. pp.156. [1000.]
iii. III. 1938. 1940. pp.164. [1000.]
iii. IV. 1939. 1941. pp.136. [750.]
iii. V. 1940. 1946. pp.238. [1500.]
iv. I. 1941–1942. 1948. pp.207. [1400.]
iv. II. 1943–1944. 1950. pp.181. [1250.]
v. I. 1945–1946. [By O. Boone]. 1950.
     pp.202. [1400.]
v. II. 1947–1948. 1952. pp.313. [2500.]
vi. I. 1949. 1952. pp.216. [1500.]
vi. II. 1950. 1953. pp.272. [1750.]
1951. 1954. pp.3–225. [1000.]
1952. 1955. pp.3–234. [1000.]
1953. 1956. pp.3–366. [2250.]
1954. 1957. pp.3–287. [2500.]
1955. 1958. pp.3–300. [2000.]
1956. 1958. pp.3–325. [2000.]
1957. 1959. pp.3–231. [1250.]
1958. 1960. pp.[v].358. [1250.]
1959. 1961. pp.[iv].383. [1500.]

in progress.

Anthropology

Costa Rica

JORGE A. LINES, Bibliografía antropológica aborigen de Costa Rica. Facultad de letras y filosofía: San José 1943. pp.263. [1262.]

Czechoslovakia

LUDVÍK KUNZ, Česká ethnografie a folkloristika v letech 1945–1952. Československá akademie ved: Praha 1954. pp.384. [2721.]

MILAN DOKLÁDAL, Československá anthropologická bibliografie 1945–1954. Brno 1955. pp.40. [755.]

Ecuador

CARLOS MANUEL LARREA, Bibliografía cientifica del Ecuador. [Parte cuarta. Antropologia, etnografia, arqueologia]. Quito 1952. pp.[v].325–561. [1516.]

Europe

WILLIAM Z[EBINA] RIPLEY, A selected bibliography of the anthropology and ethnology of Europe. [Edited by Lindsay Swift.] Public library: Boston 1899. pp.x.160. [1500.]

   reissued *New York 1899 as a supplement to the author's* Races of Europe.

JOEL [MARTIN] HALPERN [*and others*], Bibliogra-

phy of anthropological and sociological publication on eastern Europe and the U.S.S.R. (English
language sources). University of California: Russian and east european studies center series (vol.1,
no.2): Los Angeles 1961. pp.[vii].ii.142. [2000.]*

France

G[USTAVE] CHAUVET, Stations humaines quater-
naires de la Charente. . . . Nº 1. Bibliographie et
statistique. Fouilles au Ménieux et à La Quina.
[Angoulême] 1897. pp.137. [500.]

G[USTAVE] CHAUVET, Statistique et bibliographie des sépultures du département de la Charente.
1900. pp.56. [300.]

Germany

WILHELM BLASIUS, Die anthropologische literatur Braunschweigs und der nachbargebiete mit
einschluss des ganzen Harzes. Braunschweig 1900.
pp.[ii].231. [2028.]

Guatemala

ROBERT H[AROLD] EWALD, Bibliografía comentada sobre antropología social guatemalteca 1900–
1955. [Translated by Carlos Delgado]. Seminario
de integración social guatemalteca: Guatemala
1956. pp.133. [292.]

Anthropology

India

DAVID G[OODMAN] MANDELBAUM, Materials for a bibliography of the ethnology of India. [Berkeley, Cal. 1949]. ff.[iii].220. [2300.]*

Indonesia

JOHANNES P[IETER] KLEIWEG DE ZWAAN, Anthropologische bibliographie van den Indischen archipel en van Nederlandsch West-Indië. Bureau voor de bestuurzaken der buitengewesten: Mededeelingen (vol.xxx). Batavia 1923. pp.473.

RAYMOND KENNEDY, Bibliography of indonesian peoples and cultures. Yale university: Department of anthropology: Yale anthropological studies (vol.iv): New Haven 1945. pp.212. [10,000.]*
— — Revised edition. Editors . . . Thomas W. Maretzki and H. Th. Fischer. Human relations area files: Behavior science bibliographies: 1955. pp.[ii].xxviii.320+[ii].321–663. [12,500.]*
reprinted in 1962 in a different format.

Italy

LUIGI PIGORINI, Bibliografia paleoetnologica italiana dal 1850. Parma 1871. pp.45. [350.]

LOUIS PIGORINI, Matériaux pour l'histoire de la paléoethnologie italienne. . . . Bibliographie. Parme 1874. pp.96. [600.]

Anthropology

PAOLO RICCARDI, Saggio di un catalogo bibliografico antropologico italiano. Modena 1883. pp.[v].176. [1009.]

GIAN FRANCESCO GAMURRINI, Bibliografia dell'Italia antica. . . . Parte generale. Parte prima— Le origini. Roma.

    i. Preistoria, paletnologia italiana, il paese (geologia). 1933. pp.iii–lxxv.471. [10,000.]

    ii. Paleontologia vegetale, animale, umana. — Ricerche regionali paleontologiche e paletnologiche. 1936. pp.[ii].479. [10,000.]

Japan

RENÉ SIEFFERT, Études d'ethnographie japonaise. Maison franco-japonaise: Bulletin (n.s.ii): Tōkyō 1953. pp.[i].203. [378.]

Kabylia

CAMILLE LACOSTE, Bibliographie ethnologique de la grande Kabylie. Maison des sciences de l'homme: Recherches méditerranéennes: Bibliographies (no.1): 1962. pp.3–104. [732.]

Kazakhstan

ALEKSYEI [NIKOLAEVICH] KHARUZIN, Библіографическій указатель статей, касающихся

Anthropology

этнографій киргизовъ и каракиргизовъ съ 1734 по 1891 г. Москва 1891. pp.68. [800.]

Kirghiztan

Z. L. AMITIN-SHAPIRO, Аннотированный указатель литературы по истории, археологии и этнографии Киргизии, 1750–1917. Академия наук Киргизской СССР: Институт истории: Фрунзе 1958. pp.353. [2168.]

Korea

BERT A. GEROW, Publications in japanese on korean anthropology. A bibliography of uncatalogued material in the [Takeo] Kanaseki collection, Stanford university library. [Stanford 1952]. ff.iv.18. [225.]*

Malaysia

P[ETER] SUZUKI, Critical survey of studies on the anthropology of Nias, Mentawei and Enggano. Koninklijk instituut voor taal-, lund- en volkenkunde: Bibliographical series (no.3): 's-Gravenhage 1958. pp.[vii].87. [500.]

Mexico

NICOLÁS LEÓN, Apuntes para una bibliografía

43

antropológica de México. Museo nacional de México: Sección de antropología y etnografía: México 1901. pp.18. [167.]

HERMANN BAYER, Succinta, bibliografía sistemática de etnografía y arqueología mexicanas. Universidad nacional de México: Facultad de altos estudios: México 1923. pp.40. [400.]

JUAN COMAS and SANTIAGO GENOVÉS T[ARAZAGA], La antropología física en México, 1943–1959. Inventario y programa de investigaciones. Universidad nacional autónoma de México: Instituto de historia: Cuadernos: Serie antropológica (no.10): México 1960. pp.67. [300.]

JORGE MARTÍNEZ RÍOS, Bibliografía antropológica y sociológica del estado de Oaxaca. Universidad nacional: Instituto de investigaciones sociales: México 1961. pp.154.

Mozambique

ANTONIO RITA-FERREIRA, Bibliografia etnológica de Moçambique das origens a 1954. Junta de investigações do ultramar: Lisboa 1961. pp.xiii.254.

Negro

LIST of references on the anthropology and

ethnography of the Negro race. Library of Congress: Washington 1905. ff.5. [26.]*
— Additional references. 1916. ff.3. [31.]*

H[EINRICH] A[LBERT] WIESCHHOFF, Anthropological bibliography of negro Africa. American oriental series (vol.xxiii): New Haven 1948. pp.xi.461. [18,500.]*

Netherlands

A. J. BORK-FELTKAMP, Anthropological research in the Netherlands. Koninklijke akademie van wetenschappen: Afdeeling natuurkunde: Verhandelingen (vol.xxxvii, no.3). Amsterdam 1938. pp.166. [630.]

Pacific

C[LYDE] R[OMER] H[UGHES] TAYLOR, A Pacific bibliography. Printed matter relating to the native peoples of Polynesia, Melanesia and Micronesia. Polynesian society: Memoirs (vol.xxiv): Wellington, N.Z. 1951. pp.xxix.492. [11,000.]

RECHERCHES en sciences sociales dans les îles du Pacifique. Commission du Pacifique sud: Document technique (no.20 &c.): Nouméa.

1951. . . . (no.20).
1953. . . . (no.52). pp.vi.34. [60.]

1956. .... (no.98). pp.[v].60. [120.]
1959. .... (no.127). pp.[v].70. [150.]
1961. .... (no.135). pp.[v].47. [100.]

IRWIN HOWARD, W. EDGAR VINACKE and THOMAS MARETZKI, Culture & personality in the Pacific Islands. A bibliography. Anthropological society of Hawaii: Honolulu 1963. pp.[ii].iv.110. [1000.]*

Persian gulf

HENRY FIELD, Anthropogeographical bibliography of the Persian gulf area. [*s.l.*] 1952. ff.[i].17. [250.]*

Peru

GEORGE A[MOS] DORSEY, A bibliography of the anthropology of Peru. Field columbian museum (Publication 23=Anthropological series, vol.ii, no.2): Chicago 1898. pp.[v].55–206. [1500.]

FEDERICO SCHWAB, Bibliografía etnológica de la Amazonia peruana, 1542–1942. [Comité del IV centenario del descubrimiento del río Amazonas:] Lima 1942. pp.[ii].77. [350.]

Poland

ADOLF STRZELECKI, Materyały do bibliografii

etnograficznej (1878–1894). Lwów 1901. pp.212.
[2300.]

S. J. CZARNOWSKI, Literatura przeddziejów
Polski i ziem sąsiednich słowiańskich. Polska.
przedhistoryczna: Warszawa &c. 1909. pp.[ii].149.
[1000.]

FRANCISZEK GAWEŁEK, Bibliografia ludoznaw-
stwa polskiego. Akademija umiejętności: Kraków
1914. pp.xlii.328. [7250.]

JAN ST[ANISŁAW] BYSTROŃ, Bibljografja etno-
grafji polskiej. I. Bibljoteka 'ludu słowiańskiego'
(no.1): Kraków 1929. pp.vi.160. [1701.]

HALINA BITTNER[-SZEWCZYKOWA], ANNA ŁYSZCZ
and ZOFIA STASZCZAK, Materiały do bibliografii
etnografii polskiej (1945–1955). . . . Wykaz
ksiçizek nabytych przez Polskie towarzystwo
ludoznawcze (1951–1955). . . . Wykaz zawartości
archiwum Naukowego polskiego towarzystwa
ludoznawcze (1945–1955). Archiwum etnograficzne (no.11): Wrocław [1955]. pp.62.53.28.
[3837.]

HALINA BITTNER-SZEWCZYKOWA, Materiały do
bibliografii etnografii polskiej za 1945–1954 r.
Polskie towarzsystwo ludoznawcze: Ludu (vol.

xliii, supplement): Wrocław 1958. pp.355. [3828.]

Russia

v[LADIMIR IZMAILOVICH] MEZHOV, Библіографическій указатель русской этнографической литературы (книгъ и статей) за 1860 и 1861 года. Санктпетербургъ 1864. pp.[ii]. 132. [352.]

D. K. ZELENNY, Библіографическій указатель русской этнографической литературы о внѣшнемъ бытѣ народовъ Россіи, 1700–1900 г.г. С.-Петербургъ 1913. pp.xxxix.736. [9000.]

LIST of references on the races of Russia. Library of Congress: Washington 1915. ff.5. [62.]*

F. P. SCHILLER, Literatur zur geschichte und volkskunde der deutschen kolonien in der Sowet [*sic*]-Union für die Jahre 1764–1926. Prokrowska. W. 1927. pp.67. [970.]

JOEL [MARTIN] HALPERN [*and others*], Bibliography of anthropological and sociological publication on eastern Europe and the U.S.S.R. (English language sources). University of California: Russian and east european studies center series (vol.1, no.2): Los Angeles 1961. pp.[vii].ii.142. [2000.]*

Anthropology

Spain

GABRIEL PUIG Y LARRAZ, Ensayo bibliográfico de antropología prehistórica ibérica. Madrid 1897. pp.88. [257.]

FLORENTINO L[ÓPEZ] CUEVILLAS and FIRMÍN BOUZA BREY, Bibliografía da prehistoria galega. Publicazóns do 'Seminario de estudos galegos': Seizón de prehistoria: Cruña 1927. pp.118. [220.]

Switzerland

RUDOLF MARTIN, Physische anthropologie der schweizerischen bevölkerung. — Jakob Heierli, Urgeschichte der Schweiz. Centralkommission für schweizerische landeskunde: Bibliographie der schweizerischen landeskunde (section V.2): Bern 1901. pp.iv.138. [2250.]

Tierra del Fuego

JOHN M[ONTGOMERY] COOPER, Analytical and critical bibliography of the tribes of Tierra del Fuego and adjacent territory. Bureau of american ethnology: Bulletin (no.63): Washington 1917. pp.ix.233. [750.]

United States

RICHARD G[OOCH] BEIDLEMAN, A partial, an-

notated bibliography of Colorado ethnology.
Colorado college studies (no.2): Colorado Springs
1958. pp.55. [544.]

West Indies

JOHANNES P[IETER] KLEIWEG DE ZWAAN, Anthropologische bibliographie van den Indischen archipel en van Nederlandsch West-Indië. Bureau
voor de bestuurzaken der buitengewesten. Mededeelingen (vol.xxx): Batavia. 1923. pp.473.

4. Cartography

CARLO MAGNINO and MARIO DE MANDATO,
Bibliografia per una cartoteca etnica. I. Europa
centro-orientale. Centro di documentazione
etnica: Roma 1933. pp.53. [300.]

5. 'Race'

ACHIM GERCKE, Die rasse im schrifttum. Ein
wegweiser durch das rassekundliche schrifttum ...
Bearbeitet von Rudolf Kummer. Berlin [1933].
pp.92. [700.]

ARTHUR STEIDING, Stoffverteilungsplan für den
unterricht in familienkunde, vererbungslehre,
rassenkunde, erbgesundheits- und rassenpflege
und bevölkerungspolitik. Langensalza &c. 1936.
pp.72. [1250.]

HANS RIEGELMANN, Norden in abwehr. Leipzig 1937. pp.74. [122.]

GUIDO LANDRA and GIULIO COGNI, Piccola bibliografia razziale. Ragguagli bibliografici (vol.iv): Roma [1939]. pp.3–80. [125.]

JOHN G. ILIFF, Where to read up on racism and human rights. American civil liberties union: Northern California branch: San Francisco 1954. pp.79. [1500.]

HAROLD SMITH, Apartheid in the Union of South Africa. Library association: Special subject list (no.5): 1956. pp.12. [113.]

A[LEXANDRA] M[ARGARET] MACDONALD, A contribution to a bibliography on university apartheid. University of Cape Town: School of librarianship: Cape Town 1959. pp.[iii].iv.30. [164.]★

RACE relations, with special reference to employment. International labour office: Library: Bibliographical reference list (no.102): Geneva 1963. pp.8. [150.]★

6. *Teaching*

DAVID G[OODMAN] MANDELBAUM [*and others*],

Resources for the teaching of anthropology. American anthropological association: Memoir (no.95): [Washington] 1963. pp.[v].316. [1714.]*

7. *Miscellaneous*

JEAN LEYDER, Ethnologues contemporains. . . . Paul Schebesta. Notes bio-bibliographiques. Université libre: Institut de sociologie Solvay: Bruxelles 1932. pp.8. [29.]

BIBLIOGRAPHIE sur la méthode statistique appliquée à la médecine et aux sciences anthropologiques. Liste d'ouvrages et d'articles en langue française. Faculté de médecine and Institut national de la statistique et des études économiques: 1955. pp.34. [300.]*

GEORGE E. FAY, A bibliography of fossil man. Southern state college: Department of sociology and anthropology: Magnolia, Ark. 1959 &c.*
in progress.

MARION PEARSALL, Medical behavioral science. A selected bibliography of cultural anthropology, social psychology, and sociology in medicine. [Lexington, Ky.] 1963. pp.ix.134. [3064.]*

Biology.

1. *Bibliographies*

CLARENCE J. WEST and CALLIE HULL, List of manuscript bibliographies in the biological sciences. National research council: Reprint and circular series (no.45): Washington 1923. pp.51. [450.]

ZELDA D. KNOWLES, A list of russian review papers in biology and medicine. Public health service: National institutes of health: Bethesda, Md. 1958. pp.[iii].30. [301.]*
—— Supplement. 1958. pp.[ii].5. [49.]*

V[LADIMIR] L[AZAREVICH] LEVIN, Справочное пособие по библиографии для биологов. Академия наук СССР: Отделение биологических наук: Институт цитологии: Ленинград 1960. pp.407. [1000.]

Biology

2. Periodicals

[DUNCAN STARR JOHNSON and JOHN HENDLEY BARNHART], List of biological serials in the libraries of Baltimore. Baltimore 1902. pp.41. [600.]

LIST of biological serials, exclusive of botany, in the libraries of Philadelphia. Wistar institute of anatomy and biology: Bulletin (no.2): Philadelphia 1909. pp.[ii].61. [1250.]

CHECK list of periodicals and serials in the biological and allied sciences available in the library of the university of Minnesota and its vicinity. University of Minnesota: Research publications: Bibliographical series (no.2): 1925. pp.vii.126. [2500.]

JESSIE L[OVERING] METCALF, A check list of periodicals and other serials in the biological sciences in the Detroit public library and libraries of Wayne university. Wayne university: Department of biology: Detroit 1941. pp.[ii].64. [750.]*

RUDOLPH H. GJELSNESS, MARIA THERESA CHÁVEZ and HELEN M. RANSON, Catálogo colectivo de publicaciones periódicas existentes en las bibliotecas de la ciudad de México. Sección de medicina y ciencias biológicas. Comisión impulsora y coordinadora de la investigación científica [&c.]:

Biology

México 1949. pp.xix.498. [6000.]

HARALD OSTVOLD and CAROL LEMS, Biological periodicals at Northwestern university, Evanston campus. A bibliography. [Evanston, Ill. 1949]. pp.[i].69. [1250.]*

[RALPH W. MARSH], A list of abbreviations of the titles of biological journals. Selected . . . from the "World list of scientific periodicals". Biological council: [1949]. pp.[iv].26. [1000.]

UNION list of periodicals and other serial publications in the medical and biological sciences libraries of the greater Los Angeles area. Special libraries association: Southern California chapter: Los Angeles 1951. pp.[xi].262. [3000.]*

[JOHN HENRY RICHTER and CHARLES P. DALY], Biological sciences serial publications. A world list 1950–1954. Library of Congress: Reference department: Science division: Washington 1955. pp. [viii].269. [4000.]*

ANIELA SZWEJCEROWA and ALEKSANDER SZWEJCER, Spis zagranicznych biologicznych czasopism i wydawnictw ciągłych znajdujących się w bibliotekach polskich. Materiały bibliograficzne. Państwowy instytut biologii doświadczalnej imiena M. Nenckiego: Warszawa &c. 1951. pp.iii–xxiv. 704. [5000.]

Biology

ANIELA SZWEJCEROWA and JADWIGA GROSZYŃ-
SKA, Spis polskich biologicznych czasopism i
wydawnictw ciągłych znajdujących się w bibliotekach polskich. Materiały bibliograficzne. Państwowy instytut biologii doświadczalnej im. M.
Nenckiego: Warszawa &c. 1952. pp.xvi.224.
[1000.]

3. General

JAHRESBERICHT über die fortschritte in der biologie. Erlangen.

 1843. Edited by [Carl] Canstatt and [J. G.]
  Eisenmann. pp.[ii].250. [1000.]
 1844. pp.[ii].256. [1000.]
 1845. pp.viii.272. [1000.]
 1846. pp.208. [1000.]
 1847. pp.164. [750.]
 1848. pp.[vi.180. [750.]
  [continued as:]
C. Canstatt's Jahresbericht [&c.].
 1849. Edited by [J. G.] Eisenmann. pp.[iv].
  198. [750.]
 1850. pp.[iv].147. [500.]
forms vol.i of the Jahresbericht über die fortschritte der gesammten medicin for the years shown.

ZOOLOGISCHER jahresbericht. . . . Allgemeine
biologie [1890 &c.: und entwickelungslehre].

56

Biology

Zoologische station zu Neapel: Berlin.

    1886. Referent: P. Schiemenz. pp.9. [100.]

    1887. pp.18. [75.]

    1888. pp.15. [100.]

    1889. pp.25. [125.]

    1890. Referent: Paul Mayer. pp.25. [175.]

    1891. pp.34. [175.]

    1892. pp.35. [225.]

    1893. pp.19. [150.]

    1894. pp.17. [125.]

    1895. pp.21. [125.]

    1896. pp.15. [125.]

    1897. pp.17. [150.]

    1898. pp.17. [150.]

    1899. pp.18. [150.]

    1900. pp.13. [100.]

    1901. pp.16. [150.]

    1902. pp.22. [150.]

    1903. pp.24. [150.]

    1904. pp.17. [150.]

    1905. Referenten: P. Mayer und Raymond Pearl. pp.20. [200.]

    1906. pp.23. [200.]

    1907. Referenten: P. Mayer und J. Gross. pp.18. [150.]

    1908. pp.18. [125.]

    1909. pp.20. [150.]

1910. pp.23. [175.]

1911. pp.22. [175.]

1912. pp.23. [175.]

earlier issues and a later issue were not published separately.

CATALOGUE of books in the medical and biological libraries at University college, London. 1887. pp.[ii].411. [12,500.]

L'ANNÉE biologique. Comptes rendus annuels des travaux de biologie générale. [Vols.xxv &c.: Fédération des sociétés de sciences naturelles].

 i. 1895. Edited by Yves Delage. pp.xlvii.732. [1000.]

 ii. 1896. pp.xxxv.808. [1000.]

 iii. 1897. pp.xxxv.842. [1000.]

 iv. 1898. pp.xxxi.847. [1000.]

 v. 1899–1900. pp.lxxvi.677. [1500.]

 vi. 1901. pp.lxxxiv.575. [1000.]

 vii. 1902. pp.xcii.642. [1500.]

 viii. 1903. pp.xxiv.475. [1500.]

 ix. 1904. pp.xxx.514. [2000.]

 x. 1905. pp.xv.501. [2000.]

 xi. 1906. pp.xlii.508. [2000.]

 xii. 1907. pp.xv.571. [2000.]

 xiii. 1908. pp.xxxii.517. [2000.]

 xiv. 1909. pp.xxxiv.545. [2000.]

Biology

xv. 1910. pp.xix.578. [2000.]
xvi. 1911. pp.xx.596. [2000.]
xvii. 1912. pp.xviii.695. [2500.]
xviii. 1913. pp.xx.603. [2000.]
xix. 1914. pp.xxxvi.588. [2000.]
xx. 1915. pp.lxxiii.487. [1500.]
xxi. 1916. pp.xvii.428. [1500.]
xxii. 1917. pp.lxviii.469. [1500.]
xxiii. 1918. pp.xiii.422. [1500.]
xxiv. 1919. pp.
xxv. 1920–1921. pp.xix.531. [2000.]
xxvi. 1921–1922. pp.vi.749. [2000.]
xxvii. 1922–1923. pp.203. [2500.]
xxviii. 1923–1924. pp.vi.324.23+460. [3500.]
xxix. 1924–1925. pp.viii.695+viii.809. [6000.]
xxx. 1925–1926. pp.viii.802+viii.799. [6000.]
xxxi. 1926–1927. pp.viii.753+viii.884. [6000.]
xxxii. 1927–1928. pp.vii.940+vii.858. [7000.]
xxxiii. 1928–1929. pp.vii.936+vii.781. [7000.]
xxxiv. 1930. pp.vii.920+vii.811. [7000.]
xxxv. 1931. pp.vii.1102+vii.927. [8000.]
xxxvi. 1932. pp.vii.920+vii.843. [7000.]
xxxvii. 1933. pp.vii.795+689. [7000.]
xxxviii. 1934. pp.vii.855+vii.649. [7000.]
xxxix. 1935. pp.vii.905+vii.712. [7000.]
xl. 1936. pp.vii.852+vii.695. [7000.]
xli. 1937. pp.vii.758+vii.640. [7000.]

xlii. 1938. pp.vii.681+vii.638. [7000.]
xliii. 1939. pp.vii.668+vii.564. [6000.]
xliv. 1940. pp.vi.795+vi.451. [6000.]
xlv. 1941. pp.vi.803. [4000.]
xlvi. 1942. pp.vi.626. [3000.]
xlvii. 1943. pp.vi.519. [2500.]
xlviii. 1944. pp.472. [2500.]

thereafter the publication took on a more general character.

GENERAL biology. International catalogue of scientific literature (section L): Royal society.

i. 1903. pp.xiv.144. [982.]
ii. 1904. pp.viii.120. [689.]
iii. 1905. pp.viii.128. [822.]
iv. 1906. pp.viii.148. [1008.]
v. 1907. pp.viii.141. [972.]
vi. 1908. pp.viii.154. [1106.]
vii. 1909. pp.viii.158. [1231.]
viii. 1910. pp.viii.138. [1139.]

nos.6001–6810 were duplicated.

ix. 1912. pp.viii.117. [974.]
x. 1913. pp.viii.138. [1390.]
xi. 1913. pp.viii.130. [1284.]
xii. 1914. pp.viii.111. [1058.]
xiii. 1916. pp.viii.93. [864.]
xiv. 1919. pp.viii.105. [943.]

no more published.

Biology

OTTO E. A. HJELT, Sveriges biologiska disputations- och program-litteratur, 1700-vårterminen 1910. Bidrag till kännedom af Finlands natur och folk [vol.lxx, no.1]: Helsingfors 1911. pp.viii.210. [2250.]

JAHRESBERICHT über die wissenschaftliche biologie, zugleich bibliographisches jahresregister der berichte über die wissenschaftliche biologie. Berlin.
    i. 1926. Herausgegeben von Tibor Péterfi. pp.xii.627. [7600.]
    ii. 1927. pp.xii.624. [8774.]
    iii. 1928. pp.xi.684. [11,274.]
    iv. 1929. pp.viii.592. [10,600.]
    v. 1930. pp.viii.499. [9420.]
    vi. 1931. pp.viii.561. [10,565.]
no more published.

BIOLOGICAL abstracts. A comprehensive abstracting and indexing journal of the world's literature in theoretical and applied biology, exclusive of clinical medicine. In its departments dealing with ... bacteriology and botany the journal represents a continuation of Abstracts of Bacteriology and Botanical abstracts. Union of american biological societies: Philadelphia.
    i. 1927. pp.vii.1590. [14,506.]
    ii. 1928. pp.vii.2400. [20,124.]

Biology

iii. 1929. pp.vi.2704. [23,071.]
iv. 1930. pp.vi.3531. [30,000.]
v. 1931. pp.vi.3750. [30,000.]
vi. 1932. pp.vi.3426. [26,158.]
vii. 1933. pp.vi.2985. [22,843.]
viii. 1934. pp.vi.3162. [21,469.]
ix. 1935. pp.vi.2996. [20,660.]
x. 1936. pp.vi.2988. [22,787.]
xi. 1937. pp.iv.2672. [20,074.]
xii. 1938. pp.vi.1903. [17,124.]
xiii. 1939. pp.vi.2154. [18,108.]
xiv. 1940. pp.vi.1962. [17,090.]
xv. 1941. pp.vii.2764. [24,811.]
xvi. 1942. pp.vii.2798. [23,491.]
xvii. 1943. pp.viii.2938. [25,999.]
xviii. 1944. pp.viii.2873. [23,369.]
xix. 1945. pp.viii.2994. [23,498.]
xx. 1946. pp.viii.2752. [21,782.]
xxi. 1947. pp.viii.3108. [26,660.]
xxii. 1948. pp.viii.3191. [26,265.]
xxiii. 1949. pp.iv.3761. [30,725.]
xxiv. 1950. pp.3688. [38,371.]
xxv. 1951. pp.4110. [38,422.]
xxvi. 1952. pp.3356. [37,357.]
xxvii. 1953. pp.3812. [33,498.]
xxviii. 1954. pp.3572. [30,037.]
xxix. 1955. pp.iii.3672. [30,058.]

xxx. 1956. pp.iii.4447. [36,080.]

xxxi. 1957. pp.4488. [40,061.]

xxxii. 1958. pp.4648. [42,575.]

xxxiii–xxxiv. 1959. pp.5294+1745. [62,559.]

xxxv. 1960. pp.8109. [72,532.]

xxxvi. 1961. pp.10,479. [87,022.]

xxxvii–xxxviii. 1962. pp.2574 + 2735. [50,044.]

xl–xliv. 1963. pp.2688+2758+2631+2808 +2793. [122,197.]

in progress.

WILLIAM JOHN DAKIN, What to read on biology. Public libraries: Leeds 1928. pp.32. [150.]

BIBLIOGRAFIA italiana . . . Gruppo A *bis.* Biologia. Consiglio nazionale delle ricerche: Roma.

1932. 1933. pp.[v].234. [2556.]

1933. 1934. pp.[v].380. [4775.]

1934. [1935]. pp.v.722. [10,250.]

in progress.

CURRENT titles from biological journals. A register of selected tables of contents. Vol.I, no.I[–3]. Chicago 1937. pp.76.73.76. [2500.]

no more published.

C[YRIL] E[DWARD] LUCAS, A select bibliography

on biology. Association of tutors in adult education: Leicester 1937. pp.39. [447.]

READERS' guide to books on biology, botany and zoology. Library association: County libraries section [no.17]: 1938. pp.iv].27. [600.]

CRISÓFORO VEGA, Bibliografía de los trabajos del Instituto de biología, 1930 a 1937. Universidad de México: Chapultepec 1939. pp.64. [324.]

L. YA. BLYACHER, *ed.* Биология. Государственная библиотека СССР им. В. И. Ленина: Книга о лучших книгах: [Moscow] 1941. pp.160. [200.]

INTERNATIONAL abstracts of biological sciences.*
iv. July–December 1956. pp.530.150. [7530.]
v–vii. 1957. pp.380.158+517.205+401.128. [17,940.]
viii–xi. 1958. pp.488.137+506.120+423.135 +540.148. [23,385.]
xii–xv. 1959. pp.561.136+584.145+544.151 +536.197. [25,315.]
xvi–xix. 1960. pp.480.195+513.160+552.179 +579. [24,400.]
xx–xxiii. 1961. pp.534.186+534.22.23+ 594.24. [18,209.]

xxiv–xxvii. 1962. pp.644.26+574.24+670. 24+700.28. [25,808.]

xxviii–xxxi. 1963. pp.614.24+614.26+654. 28+706.28. [25,490.]

in progress; vols.i–iii were published as British abstracts of medical sciences, *which is entered under* Medicine, *below.*

ERWIN BÜNNING, ALFRED KUHN [*and others*], Biology. Office of military government for Germany: Field information agencies technical: Fiat review of german science, 1939–1946: Wiesbaden 1948. pp.[viii].213 + [iv].208 + [iv].236 + [iv].170. [1500.]

the text is in german.

J[OHN]MURRAY SPEIRS, Bibliography of canadian biological publications for 1946. Department of lands and forests: Division of research: Biological bulletin (no.3): Ontario 1949. pp.91. [1055.]

S. S. LEVINA and S. G. MARCHENKO, Как возникла и развивалась жизнь на земле. Библиотека СССР имени В. И. Ленина, Москва 1950. pp.54. [100.]

СИСТЕМАТИЧЕСКИЙ указатель статей в иностранных журналах. Биологические науки. Всесоюзная государственная библиотека иностранной литературы: Москва.

 i. 1948. 1950. pp.514. [4428.]

ПРЕДВАРИТЕЛЬНЫЙ каталог книг по биологии. Институт по изучению истории и культуры СССР: Сообщения библиотеки института (no.3): Мюнхен 1954. pp.26. [200.]*

РЕФЕРАТИВНЫЙ журнал. Биология. Академия наук СССР: Институт научной информации: Москва.

 1954. pp.382.x +329.xi + 274.xiv + 301.
 xiv + 311.xvi + 278.xv + 312.xviii +
 335.xx. [15,937.]
 1955. pp.348.xx + 347.xx + 369.xxii + 350.
 xx + 344.xxi + 363.xxii + 314.xix +
 311.xx + 312.xx + 320.xxi + 316.xxiv +
 332.xxiv + 286.xxii + 314.xxiii + 320.
 xxiv + 307.xxiv + 350.xxv + 366.xxvii
 + 371.xxviii + 385.xxx + 406.xxx +
 411.xxxii + 462.xxxvii. [69,598.]
 — Авторский указатель за 1954–1955 г.
 pp.2175.
 1956. pp.447.xx + 431.xx + 398.xviii + 407.
 xviii + 413.xx + 435.xx + 452.xx + 439.
 xx + 452.xxi + 480 + 476 + 480 + 496 +
 495 + 493 + 496 + 496 + 513 + 504 +
 508 + 553 + 581 + 576 + 568. [107,610.]
 1957. pp.599 + 552 + 488 + 464 + 548 +

587 + 557 + 565 + 561 + 576 + 588 +
564 + 557 + 504 + 504 + 509 + 501 +
480 + 476 + 460 + 448 + 392 + 388 +
400. [103,445.]

—Предметный указатель за 1956–1957 гг.
1961. pp.1268.

1958. pp.508 + 500 + 532 + 512 + 528 +
520 + 524 + 501 + 508 + 521 + 529 +
527 + 517 + 513 + 524 + 512 + 525 +
504 + 508 + 476 + 492 + 481 + 468 +
392. [107,502.]

1959. pp.547 + 528 + 528 + 525 + 528 +
536 + 532 + 545 + 532 + 532 + 526 +
529 + 508 + 532 + 508 + 504 + 456 +
480 + 489 + 497 + 496 + 500 + 528 +
641. [104,465.]

1960. pp.528 + 532 + 532 + 552 + 544 +
560 + 528 + 540 + 551 + 556 + 544 +
540 + 548 + 536 + 532 + 524 + 528 +
512 + 450 + 488 + 560 + 616 + 708 +
528 + 700. [119,971.]

1961. pp.[394] + [560] + [584] + [560] +
[568] + [558] + [560] + [558] + [563] +
[584] + [568] + [560] + [554] + [558] +
[598] + [560] + [562] + [522] + [494] +
[543] + [552] + [576] + [506] + [624] +
644 + 831. [125,000.]

1962. pp.[600] + [544] + [630] + [574] +
[592] + [578] + [598] + [550] + [606] +
[550] + [604] + [603] + [648] + [624] +
[654] + [528] + [644] + [592] + [642] +
[616]+

in progress.

[MIECZYSŁAW BOGUCKI *and others*], Polske bibliografia analityczna. Biologia. Polska akademia nauk: Ośrodek bibliografii i dokumentacij naukowej: Warszawa.

    i. 1955. pp.67. [121.]
    ii. 1956. pp.96.121. [505.]
    iii. 1957. pp.109. [275.]
no more published.

INHALTSVERZEICHNISSE sowjetischer fachzeitschriften. Reihe III B. Chemie, biologie. Zentralstelle für wissenschaftliche literatur: Berlin.

    v. 1956. pp.1578. [4500.]
    vi. 1957. pp.2144.181. [6000.]
    vii. 1958. pp.2281. [6000.]
    viii. 1959. pp.2476. [6500.]
    ix. 1960.
    x. 1961.
    xi. 1962.
    xii. 1963.

in progress; the earlier and parallel volumes are entered under Science, above.

БИОФИЗИКА, биохимия, физиология, микробиология. Всесоюзная государственная библиотека иностранной литературы: Москва.

    1956. pp.[764]. [14,000.]
    1957. pp.[934]. [17,500.]
    1958. pp.[1286]. [25,000.]
in progress.

НОВЫЕ КНИГИ за рубежом. Серия В. Биология, медицина, сельское хозяйство. Критико-библиографический бюллетень. Москва.

    [i]. 1957. pp.136.120.136.128.104.111. [2000.]
    ii. 1958. pp.
    iii. 1959 pp.[1140]. [3000.]
    iv. 1960. pp.[1166]. [3000.]
    v. 1961. pp.[1200]. [3000.]
in progress.

ABSTRACTS of bulgarian scientific literature. Biology and medicine. Bulgarian academy of sciences: Department for scientific and technical information and documentation: Sofia.★

    [i]. 1958. pp.[ix].184+[ii].149+[ii].244. [791.]
    ii. 1959. pp.[ii].172+

QUARTERLY review of scientific publications of

the polish academy of sciences, the Ossolineum and the polish scientific publishers. Series B. Biological sciences. Polish academy of sciences: Distribution centre for scientific publications: Warsaw.

1958.

1959. pp.59.42.40.39.16. [432.]

1960.

1961.

from 1962 the three series were consolidated; see Science.

DAGMAR KLESKĚNOVÁ, Metodika biologie. Bibliografie knižních publikací a časopiseckých článků. Státni pedagogická knihovna: Publikace (no.82): Brně 1960. pp.[v].119. [1167.]*

JANE CLAPP, Museum publications. Part II. Publications in biological and earth sciences. 1962. pp. 610. [9231.]*

CATALOGUE collectif des livres français de médecine et de biologie, 1952–1962. Cercle de la librairie: [1962]. pp.xii.176. [2000.]

ÍNDICE bibliográfico del Centro de investigación y de estudios avanzados. Sección 5ª. . . . 1. Biología. Instituto politécnico nacional: México 1962 &c.

in progress.

Biology

4. *Miscellaneous*

[LOUIS JOSEPH REMY SAINT-LOUP], Notice résumée sur les travaux et thèses biologiques de m. Remy Saint-Loup. Besançon [printed] 1903. pp. [iii].89. [37.]

LIST of references on microbiology. Library of Congress: Washington 1917. ff.2. [19.]*

RAPHAEL ED[UARD] LIESEGANG, Biologische kolloidchemie. Wissenschaftliche forschungsberichte: Naturwissenschaftliche reihe (vol.xix): Dresden &c. 1928. pp.xii.128. [600.]

M. G. ADAMS, A short book list of evolutionary biology. Tutors' association: Bibliographical series (no.3): [1933]. pp.12. [150.]

BIBLIOGRAPHIA biotheoretica. Universiteit te Leiden: Prof. dr. Jan van der Hoeven stichting voor theoretische biologie van dier en mensch: Geschriften (ser. C): Leiden.

  1925–1929. [By J. H. Diemer and C. J. van der Klaauw]. 1938. pp.xi.236. [4802.]
  1930–1934. 1941. pp.vii.310. [6076.]
  1935–1939. 1940–1942. pp.371. [7170.]
  1940–1944. pp.[ix].342. [6187.]
  1945–1949. 1954. pp.viii.447. [8986.]

Biology

1950–1954. 1956. pp.214. [3988.]
1955–1959. 1962–   . pp.

[RUSSELL B. STEVENS], Biological education. A partial bibliography. National research council: Publication (no.518): Washington 1957. pp.iv. 159. [4000.]*

J. A. MCCORMICK, Isotope techniques in biological sciences. A selected list of references. United States atomic energy commission: Technical information service extension: Oak Ridge, Tenn. 1958. pp.44. [807.]*

RICHARD S. CUTTER, Biological effects of non-ionizing radiation on humans and higher animals. Selected references in english, 1916–1957. National library of medicine: Washington 1958. pp. [iii].10. [118.]*

J. A. MCCORMICK, Utilization of radioisotopes in physical and biological sciences— general topics. A literature search. Atomic energy commission: Technical information service: Oak Ridge, Tenn. 1959. pp.30. [492.]*

ASBJØRN FJELD-ANDERSEN, Oversiktskatalog 1 over bøker, periodiske skrifter og særtrykk ved Hvalfangstmuseets bibliotek. Sandefjord 1961. pp.[ii].v.264. [4000.]

Biology

KATYE M. GIBBS, Bionics and related research. Armed services technical information agency: Arlington, Va. 1963. pp.viii.177. [1250.]*

CHEMICAL and biological warfare (USSR). Bibliography. Library of Congress: Aerospace information division: Report (B-63-52): [Washington] 1963. ff.[ii].34. [278.]*

Botany.

Botany

1. *Bibliographies*

BIBLIOGRAPHIES on pure and applied botany and related subjects. Science library: Bibliographical series (no.144): 1934. ff.[i].7. [163.]*

2. *Periodicals*

P. VAN AERDSCHOT, Catalogue de la bibliothèque collective réunie au Jardin botanique de l'état à Bruxelles. . . . 1. Publications périodiques ou occasionnelles d'académies, de jardins et d'instituts botaniques et de sociétés savantes. Jardin botanique: Bulletin (vol.iii): Bruxelles 1911. pp.[iii]. xxxiii.252. [1750.]

EDITH WYCOFF, Catalogue of the periodical literature in the Lloyd library. Lloyd library: Bibliographical contributions (no.1): Cincinnati 1911. pp.[ii].80. [1500.]
—— [second edition]. . . . (vol.ii, no.1): 1914. pp.[ii].123. [2000.]

LIJST von botanische tijdschriften in de universiteitsbibliotheek, in het Botanisch laboratorium en -museum, in het Phytopathologisch laboratorium

'Willie Commelin Scholten' (te Baarn), zoomede op enkele andere plaatsen te Utrecht. Rijksuniversiteit: Bibliotheek: Utrecht 1935. pp.[ii].38. [400.]

WALTER H[ARRIS] AIKEN and SIGMUND WALDBOTT, Catalogue of the periodical literature in the Lloyd library. Lloyd library and museum: Bulletin (no.34): Cincinnati 1936. pp.[iv].103. [2500.]

A LIST of periodicals dealing wholly or mainly with phytopathology. Science library: Bibliographical series (no.239): 1936. ff.4. [87.]*

[R. GUINEY], list of periodicals. University of Oxford: Department of botany: Library: Oxford 1944. pp.20. [200.]

LIST of periodicals held in the library. University of Oxford: Department of botany: Library: [Oxford] 1961. ff.[i].22. [400.]*

3. *History*

FRANCESCO TORNABENE, Ricerche bibliografiche sulle opere botaniche del secolo decimoquinto. Catania 1840. pp.91.

[TH.] CHABOISSEAU, Notes de bibliographie botanique. 1872. pp.27. [100.]

CAROLINE HUBBARD BAILEY, The romantic and

historic background of agriculture and plant study. Public library: Bulletin (no.175): Riverside, Cal. 1921. pp.39. [200.]

JANE QUINBY, Catalogue of botanical books in the collection of Rachel McMasters Miller Hunt. Hunt botanical library: Pittsburgh.

    i. Printed books 1477–1700, with several manuscripts of the 12th, 15th, 16th & 17th centuries. 1958. pp.lxxxiv.519. [461.]

4. *General*

CASPARIUS BAUHINUS, Φυτοπιναξ seu envmeratio plantarvm ab herbarijs nostro seculo descriptarum cum earum differentijs. Basileæ 1569. pp. [xliv].669.[xxii]. [6000.]

CASPARUS BAUHINUS, Πιναξ theatri botanici . . . sive index in Theophrasti, Dioscoridis, Plinii et botanicorvm qui à seculo scripserunt opera: plantarum sec millivm ab ipsis exhibitarvm nomina cvm earundem synonymis & differentiis . . . proponens. Basileæ Helvet. 1623. pp.[xxiv].522[xxiii]. [6000.]

copies in the Bibliothèque nationale and the British museum contain ms. notes, the former by Jacques Barrelier.

— — Opus . . . ad autoris autographum recensitum. 1671. pp.[xxiv].518.[xxi]. [6000.]

copies in the Bibliothèque nationale and the British museum contain ms. notes, the former by Sébastien Vaillant.

GIOVANNI ANTONIO BUMALDO [*pseud.* OVIDIO MONTALBANI], Bibliotheca botanica, sev. hervarijstarum scriptorum promota synodia. Bononię 1657. pp.189. [750.]

— — Nunc iterùm edita. Hagæ-Comitum 1740. pp.66. [400.]

issued as the second part of J. F. Seguier's Bibliotheca botanica, 1740; and reissued Lugduni Batavorum 1760.

CAROLUS LINNAEUS [CARL VON LINNÉ], Bibliotheca botanica recensens libros plus quam mille de plantis huc usque editos. Amstelodami 1736. pp.[xvi].153.[xix].36. [1000.]

— — Editio nova. [Edited by Michael Gottlieb Agnethler]. Halæ Salicæ 1747. pp.[xiv].124.[viii]. [1250.]

there are no pp.83–84.

— — Editio altera. Amstelodami 1751. pp.[xvi]. 220.[xiii]. [2000.]

the Bibliotheca was also issued with Linné's Fundamenta botanica; editions dated 1742 and 1753 are sometimes referred to, but appear to be ghosts; see also Augustinus Loo, 1759, below.

JEAN FRANÇOIS SEGUIER, Bibliotheca botanica, sive catalogus auctorum et librorum omnium qui de re botanica, de medicamentis ex vegetabilibus paratis, de re rustica et de horticultura tractant. . . . Accessit Bibliotheca botanica Jo. Ant. Bumaldi, seu potius Ovidii Montalbani. Hagæ-Comitum 1740. pp.16.450+66. [5000.]

—— Supplementum. [Veronæ 1745]. pp.79. [500.]

forms part of Seguier's Plantæ veronenses.

—— Auctuarium . . . prolatum a Laur. Theod. Gronovio. Lugduni Batavorum 1760. pp.[iv].40+ [ii].41–65.[vii]. [500.]

forms a supplement to a reissue of the main work.

CHRISTOPH JACOB TREW, Librorvm botanicorvm catalogi dvo, qvorvm prior recentiores qvosdam posterior plerosqve antiqvos ad annvm MDL vsqve excvsos, ad dvctvm propriae collectionis breviter recenset conscripti a d. Christophoro Iacobo Trew. Norimbergae 1752. pp.[56]. [1000.]

—— Catalogvs tertivs. 1757. pp.[ii].55–80. [1000.]

25 copies privately printed; the British museum copy contains ms. corrections by the author.

AUGUSTINUS LOO, Auctores botanici, in disserta-
tione propositi. [Carl von Linné *praes.*] Upsaliæ
1759. pp.[ii].20. [400.]

this is in effect a continuation of Linné's Bibliotheca *entered above; reprinted in Linné's* Amoeni-
tates academicae (*1760*), *v.* XCI, *and so in the later
editions of this work; and in his* Systema plantarum
Europae (*1785*), *i.4.*

[BARON OTTO VON MÜNCHHAUSEN], Des haus-
vaters zweyten theils erstes stück. Inhalt: des haus-
vaters botanische, physikalische und oekonomische bibliothek. Hannover 1765–1766. pp.[ii].
xxxvi.[xii].367+369–832.[ciii]. [4000.]

[BARON] ALBRECHT VON HALLER, Bibliotheca
botanica, qua scripta ad rem herbariam facienta a
rerum initiis recensentur. Tomus I. Tempora ante
Tournefortium. [II. A Tournefortio ad nostra
tempora]. Tiguri 1771–1772. pp.xvi.654+[ii].785.
[15,000.]

also issued with a London imprint.

— — Christophori Theophili de Mvrr Adnota-
tiones ad Bibliothecas Hallerianas botanicam, anatomicam, chirvrgicam et medicinae practicae.
Erlangae 1805. pp.[iii].67. [300.]

— — Bibliotheca botanica. Auctore Alberto de
Haller. Index emendatus. Perfecit J[ens] Christian

Bay. Societas bernensis rervm natvrae peritorvm Bernae 1908. pp.[v].57.

E[RNEST] G[OTTFRIED] BALDINGER, Ueber literargeschichte der theoretischen und praktischen botanik. Marburg 1794. pp.117. [500.]

J[EREMIAS] D[AVID] REUSS, Repertorium commentationum a Societatibus litterariis editarum. . . . Tom II. Botanica et mineralogia. Gottingae 1802. pp.viii.288.viii.289–604. [Botany: 2500.]

J. A. SCHULTES, Grundriss einer geschichte und literatur der botanik, von Theophrastus Eresios bis auf die neuesten zeiten. Wien 1817. pp.xvi.413. [2500.]

—— Vollständiges register, von J. Schultes. München 1871. pp.60.

J[OHANNES] E. WIKSTRÖM, Öfversigt af botaniska arbeten och upptäckter. Kongliga svenska vetenskaps academien: Stockholm.

 1825. pp.[ii].161. [125.]
  [*continued as:*]
Årsberättelse om framstegen uti botanik.
 1826. pp.[ii].viii.302. [200.]
 1827. pp.xii.372. [250.]
  [*continued as:*]
Års-berättelse om botaniska arbeten och upptäckter.

1828. pp.xii.269. [250.]
1829. pp.[ii].viii.185. [200.]
1830. pp.[ii].viii.299. [250.]
1831. pp.[ii].x.344. [250.]
1832. pp.x.292. [250.]
1833. pp.[ii].xi.265. [250.]
1834. pp.[ii].xii.185. [300.]
1835. pp.xvi.567. [400.]
1836. pp.[ii].xv.494. [400.]
1837. pp.xvi.612. [400.]
1838. pp.xvi.644. [400.]
1839–1842. pp.xxxix.823. [1000.]
1843–1844. pp.xxiv.429. [500.]
earlier issues formed part of the Academy's Årsberättelse.

— Jahresbericht der Königl. schwedischen akademie der wissenschaften über die fortschritte der
botanik. Breslau.

   i. 1820–1825. pp.230. [730.]
   ii. 1826–1827. pp.x.285. [300.]
   iii. 1828. pp.viii.128. [170.]
   iv. 1829. pp.100. [150.]
   v. 1830. pp.viii.166. [240.]
   [vi]. 1831. pp.xiv.203. [250.]
   [vii]. 1832. pp.viii.186. [270.]
   [viii]. 1833. pp.x.226. [250.]
   [ix]. 1834. pp.xii.232. [50.]

[x]. 1835. pp.xiv.426. [75.]
[xi]. 1836. pp.viii.362. [70.]
[xii]. 1837. pp.viii.435. [85.]
[xiii]. 1838. pp.532. [105.]

FRIEDRICH VON MILTITZ, Handbuch der botanischen literatur für botaniker, bibliothekare, buch-
händler und auctionatoren. Berlin 1829. pp.viii.
coll.544. [5000.]

JOHANN HEINRICH DIERBACH, Repertorium bota-
nicum, oder versuch einer systematischen darstellung der neuesten leistungen im ganzen umfange
der pflanzenkunde. Lemgo 1831. pp.xi.268.
[3500.]

M[ARCUS] S[ALOMON] KRÜGER, Bibliographia
botanica. Handbuch der botanischen literatur.
Berlin 1841. pp.vi.464. [10,000.]

REPORTS on the progress of zoology and botany.
1841, 1842. Ray society: Edinburgh [printed] 1845.
pp.viii.43.348.104.xix. [botany: 100.]

G[EORG] A[UGUST] PRITZEL, Thesaurus literaturae
botanicae omnium gentium. Lipsiae [1847–]1851.
pp.viii.548. [11,906.]
—— Additamenta . . . composuit Ernestus
Amandus Zuchold. Lipsiae 1853. pp.60. [500.]

— — Additamenta. . . . Index librorum botanicorum bibliothecae Horti imperialis botanici petropolitani quorum inscriptiones in G.A. Pritzelii Thesauro . . . et in Additamentis . . . ab Ernesto Amando Zuchold editis desiderantur. Collegit . . . Ernestus de Berg. Halis 1859. pp.40. [300.]

— — — Index II. Petropoli 1862. pp.21. [150.]

— — — Index III. Petropoli 1864. pp.69. [500.]

— — [new edition]. Lipsiae 1872[–1877]. pp. [viii.576. [10,871.]

a facsimile of this edition was issued Milano [1950].

A[UGUST HEINRICH RUDOLF] GRISEBACH, Bericht über die leistungen in der systematischen botanik während des jahres 1846. Berlin 1849. pp.[ii].26. [150.]

see also section 7 below.

CHARLES GIRARD, American zoological, botanical, and geological bibliography, for the year 1851. [*s.l.* 1852]. pp.19. [300.]

ERNESTUS DE [ERNST VON] BERG, Catalogus systematicus bibliothecae Horti imperialis botanici petropolitani. Petropoli 1852. pp.xvi.515. [6500.]

— — Editio nova. Curavit Ferdinandus ab [Ferdinand Gottfeld Theobald Maximilian von] Herder. 1886. pp.xi.510. [15,000.]

— — — [supplement].1886–1898. Curavit Joan-

nes [Johannes Christof] Klinge. 1899. pp.[ii].vii.
253. [4000.]

REPERTORIUM der periodischen botanischen literatur. Als beiblatt zur Flora. Regensburg.

 i. 1864. pp.[iv].126. [917.]
 ii. 1865. pp.[iv].96. [973.]
 iii. 1866. pp.[iv].90. [958].
 iv. 1867.
 v. 1868. pp.[ii].64. [667.]
 vi. 1869. pp.[ii].70. [792.]
 vii. 1870.
 viii. 1871. pp.[ii].53. [624.]

CAROLUS [KARL] MUELLER, Addenda ad littera-
turam botanicam annorum 1856–1866. Walpers,
Annales botanices systematicae (vol.vii): Lipsiae
1868. pp.viii.960. [15,000.]

REPERTORIUM annuum literaturae botanicae
periodicae. Harlemi.

 i. 1872. By J. A. van Bemmelen. 1873. pp.[iii].
  xvi.223. [2500.]
 ii. 1873. By G[eorg] C[arl] W[ilhelm] Bohnensieg und W[illiam] Burck. 1876. pp.
  [iii].xx.200. [2500.]
 iii. 1874. 1877. pp.[iii].xxii.271. [4000.]
 iv. 1875. 1878. pp.[iii].xxiv.283. [4000.]

Botany

v. 1876. 1879. pp.[iii].xxiv.328. [4750.]

vi. 1877. By G. C. W. Bohnensieg. 1881. pp. [iii].xxvi.420. [6000.]

vii. 1878. 1883. pp.xxx.522. [8000.]

viii. 1879. [1884–]1886. pp.xxxii.741. [9750.] *no more published.*

[CHARLES JACQUES] ÉDOUARD MORREN, Mémorandum des travaux de botanique et de physiologie végétale qui ont été publiés par l'Académie royale des sciences, des lettres et des beaux-arts de Belgique . . . (1772–1871). Bruxelles 1872. pp.[iii]. ii.96. [393.]

BOTANISCHER jahresbericht. Systematisch geordnetes repertorium der botanischen literatur aller länder. Berlin.

i. 1873. Herausgegeben von Leopold Just. pp.xxxvi.748. [1250.]

ii. 1874. pp.viii.1300. [2000.]

iii. 1875. pp.xiv.1148. [1750.]

iv. 1876. pp.xiii.1534. [2500.]

v. 1877. pp.x.1103. [1750.]

vi. 1878. pp.vi.632+v.1399. [4000.]

vii. 1879. pp.v.603+v.882. [3500.]

viii. 1880. pp.iv.593+iv.1024. [4000.]

ix. 1881. pp.iv.562+iv.1191. [5000.]

x. 1882. pp.vi.563+vi.868. [5000.]

87

[*continued as:*]

Just's Botanischer jahresbericht.

xi. 1883. Herausgegeben von E. Koehne und [Hermann] Th[eodor] Geyler. pp.iv.641+ vi.1126. [5000.]

xii. 1884. pp.iv.688+viii.750. [5000.]

xiii. 1885. pp.viii.851+viii.785. [5000.]

xiv. 1886. pp.viii.947+viii.666. [5500.]

xv. 1887. Herausgegeben von E. Koehne. pp. x.603+viii.901. [5500.]

xvi. 1888. pp.viii.789+viii.627. [5000.]

xvii. 1889. pp.viii.733+viii.587. [5500.]

xviii. 1890. pp.vi.752+x.663. [6000.]

xix. 1891. pp.vii.623+x.612. [5500.]

xx. 1892. pp.vii.614+x.622. [6000.]

xxi. 1893. pp.vii.584+x.694. [5500.]

xxii. 1894. pp.vii.506+x.614. [6000.]

xxiii. 1895. pp.vii.439+x.630. [6000.]

xxiv. 1896. pp.vii.488+ix.648. [6500.]

xxv. 1897. pp.viii.538+ix.681. [6000.]

xxvi. 1898. Herausgegeben von K[arl] Schumann. pp.vii.664+ix.714. [7500.]

xxvii. 1899. pp.vii.546+x.656. [5500.]

xxviii. 1900. pp.vii.507+x.689. [6000.]

xxix. 1901. pp.vii.584+x.1024. [7500.]

xxx. 1902. pp.vii.714+vii.1214. [10,000.]

xxxi. 1903. Herausgegeben von F[riedrich]

Botany

Fedde. pp.xi.892+ix.1326. [11,000.]

xxxii. 1904. pp.[iii].ii.886+xi.1630. [12,000.]

xxxiii. 1905. pp.viii.892+v.598+ix.1213. [12,000.]

xxxiv. 1906. pp.vi.630+vii.702+viii.1364. [14,000.]

xxxv. 1907. pp.xii.910+viii.764+viii.1044. [13,000.]

xxxvi. 1908. pp.viii.862+vii.1000+viii.982. [15,000.]

xxxvii. 1909. pp.ix.1126+viii.1414. [1650.]

xxxviii. 1910. pp.viii.1388+x.1671+[ii].378. [18,000.]

xxxix. 1911. 1913–1917. pp.viii.1333+x. 1550. [19,500.]

xl. 1912.

xli. 1913.

LUDOVICUS [LOUIS GEORG CARL] PFEIFFER, Nomenclator botanicus. Nominum ad finem 1858 publici juris factorum ... enumeratio alphabetica. Casselis[1872–]1873–1874.pp.[vi.]1876+[ii].1698. [200,000.]

GEORGE LINCOLN GOODALE, The floras of different countries. Library of Harvard university: Bibliographical contributions (no.9): Cambridge, Mass. 1879. pp.12. [400.]

Botany

WILLIAM E[DWARD] A[RMYTAGE] AXON, Botanical books in the Manchester free library. Oldham 1879. pp.22. [250.]

BOTANISCHES centralblatt. Referirendes organ für das gesammtgebiet der botanik des in- und auslandes. Cassel. [1902–1919: Association internationale des botanistes, Leiden (1906 &c.: Jena)].

 i. 1880. Herausgegeben . . . von Oscar Uhlworm. pp.[ii].lvi.800 + [ii].801–1664. [7500.]

 ii. 1881. Herausgegeben von . . . O. Uhlworm [parts 3 &c.: und W[ilhelm] J[ulius] Behrens]. pp.[ii].xviii.410 + [ii].444 + [ii].xvi. 400 + [ii].xvi.419. [7500.]

 iii. 1882. pp.[ii].xiii.472 + [ii].xv.472 + [ii]. xiii.468 + [ii].xvi.438. [10,000.]

 iv. 1883. pp.[ii].xiii.440 + [ii].xiv.400 + [ii]. xii.415 + [ii].xiv.400. [7500.]

 v. 1884. pp.[ii].xi.418 + [ii].xiii.400 + [ii].xiii. 400 + [ii].xii.393. [7500.]

 vi. 1885. pp.[ii].xiv.394 + [ii].xiii.406 + [ii]. xiii.374 + [ii].xii.388. [6000.]

 vii. 1886. pp.xvi.388 + xiii.395 + xv.360 + xiii.400. [6000.]

 viii. 1887. pp.xii.391 + xii.380 + xii.403 + xiv. 408. [6000.]

 ix. 1888. Herausgegeben . . . von O. Uhl-

worm [part 3 &c: und G. F. Kohl]. pp.xiii.
400+xiii.384+xiii.398+xiii.394. [6000.]

x. 1889. pp.xiii.424+xiii.425–864+xiii.394.
+xv.424. [6000.]

xi. 1890. pp.xv.416+xiv.400+xv.416+xiv.
432. [6000.]

xii. 1891. pp.xix.404+xviii.416+xxiii.400
+xx.384. [6000.]

xiii. 1892. pp.xiv.500[*sic*, 400]+xxiv.466+
xxiv.416+xxxii.450. [7500.]

xiv. 1893. pp.xx.416+xxvii.408+xx.400+
xxiv.400. [7500.]

xv. 1894. pp.xxiii.420+xxvi.432+xxiii.404
+xxiii.404. [7500.]

xvi. 1895. pp.xxiii.468+xxii.416+xx.400+
xxiv.448. [7500.]

xvii. 1896. pp.xxii.448+xxviii.416+xxii.400
+xvii.432. [7500.]

xviii. 1897. pp.xxiii.416+xx.416+xxiv.480
+xviii.432. [7500.]

xix. 1898. pp.xx.480+xxii.384+xviii.420+
xxi.432. [7500.]

xx. 1899. pp.xxv.448+xx.400+xvii.416+
xviii.516. [7500.]

xxi. 1900. pp.xvii.432+xxiii.416+xxviii.544
+xxii.480. [7500.]

xxii. 1901. pp.xviii.436+xxii.432+xxii.424
+xxii.448. [7500.]

xxiii. 1902. J[ohannes] P[aulus] Lotsy, chef-
redacteur.   pp.xxxix.736 + xlv.720.266.
[10,000.]

xxiv. 1903. pp.xxxviii.608+xxxix.656.336.
[10,000.]

xxv. 1904. pp.xliii.688+xl.656.288. [7500.]

xxvi.   1905.   pp.xli.672 + xxxix.656.272.
[7500.]

xxvii.   1906.   pp.xl.672 + xxxviii.672.256.
[5000.]

xxviii. 1907. pp.xli.672+xli.672.256. [5000.]

xxix. 1908. pp.xl.656+xlii.672.272. [5000.]

xxx. 1909. pp.xl.672+xlv.656.272. [5000.]

xxxi. 1910. pp.xliii.672+xlvi.656.272.[5000.]

xxxii. 1911. pp.liv.656+lii.656.288. [6000.]

xxxiii. 1912. pp.l.640+liv.688.304. [6000.]

xxxiv.   1913.   pp.xlviii.608 + lvii.688.272.
[6000.]

xxxv. 1914. pp.l.656+liii.688.256. [5000.]

xxxvi. 1915. pp.li.720+l.688.192. [4000.]

xxxvii.   1916.   pp.xlii.672 + xxxiv.592.176.
[4000.]

xxxviii.   1917.   pp.xxvi.400+xxviii.406.160.
[3000.]

xxxix. 1918. pp.xxvi.400 + xxvii.400.128. [2500.]

xl. 1919. pp.xxv.400+xxx.416.111. [2000.]
new series.

i. [1921–]1922. Herausgegeben von S. V. Simon. pp.[ii].500+[ii].125. [4000.]

ii. [1922–]1923. pp.[ii].497+[ii].128. [4000.]

iii. [1923–]1924. pp.[ii].498+[ii].129. [4000.]

iv. 1924–1925. pp.[ii].518+[ii].110. [3000.]

v. 1925. pp.[ii].516+[ii].111. [3000.]

vi. 1925–1926. pp.[ii].518+[ii].112. [3000.]

vii. 1926. pp.[ii].518+[ii].112. [3000.]

viii. 1926. pp.[ii].518+[ii].112. [3000.]

ix. [1926–]1927. pp.[ii].519+[ii].94. [2000.]

x. 1927. pp.[ii].346+[ii].56. [1500.]

— Generalregister zu den bänden 1 bis x . . . von Wilh[elm] Dörries. 1927. pp.158.

xi. 1927–1928. Herausgegeben von F. Herrig. pp.[ii].520+[ii].94. [2000.]

xii. 1928. pp.[ii].520+[ii].128. [2500.]

xiii. 1928–1929. pp.[ii].520+[ii].148. [3000.]

xiv. 1929. pp.[ii].516+[ii].148. [3000.]

xv. 1929–1930. pp.[ii].518+[ii].148. [3000.]

xvi. 1930. pp.[ii].518+[ii].148. [3000.]

xvii. 1930. pp.[ii].520+[ii].148. [3000.]

xviii. 1931. pp.[ii].520+[ii].148. [3000.]

xix. 1931. pp.[ii].518+[ii].148. [3000.]

Botany

xx. 1931–1932. pp.[ii].346+[ii].94. [1500.]
— Generalregister zu den bänden xi–xx . . .
von W. Dörries. 1932. pp.[ii].164.

xxi. 1932. pp.[ii].516+[ii].148. [3000.]
xxii. 1932–1933. pp.[ii].486+[ii].128. [2500.]
xxiii. 1933. pp.[ii].486+[ii].130. [2500.]
xxiv. 1934. pp.[ii].440+[ii].129. [2500.]
xxv. 1934–1935. pp.[ii].428+[ii].130. [2500.]
xxvi. 1935. pp.[ii].429+[ii].130. [2500.]
xxvii. 1935–1936. pp.[ii].430+[ii].128. [2500.]
xxviii. 1936–1937. pp.[ii].436 + [ii].130.
[2500.]

xxix. 1937.

xxx. 1937–1938. pp.[ii].279+[ii].94. [1500.]
— Generalregister zu den bänden xx–xxx . . .
von W. Dörrie. 1938. pp.[ii].168.

xxxi. 1938. pp.[ii].436+[ii].130. [2500.]
xxxii. 1938–1939. pp.[ii].435+[ii].130. [2500.]

BENJAMIN DAYDON JACKSON, Guide to the literature of botany. Index society: Publications (vol.
viii): 1881. pp.xl.625. [10,000.]
a facsimile was issued, New York 1964.

CATALOGUS der bibliotheek van de Nederlandsche botanische vereeniging. Nijmegen 1883. pp.
[iv].55. [500.]

C. EKAMA, Catalogue de la bibliothèque. . . .
Quatrième livraison: botanique. Fondation Teyler: Harlem 1886. pp.[iii].311–458. [900.]

HENRY TRIMEN, A classified catalogue of the
library of the Royal botanic gardens, Pérádeniya,
Ceylon. Colombo 1889. pp.[iv].28. [700.]

[J. J. BRUTEL DE LA RIVIÈRE], Catalogue de la
bibliothèque du jardin botanique de Buitenzorg.
Deuxième édition. Batavia 1894. pp.vii.371.6.
[4000.]

FRANZ VOLLMANN, Katalog der bibliothek der
Kgl. botanischen gesellschaft in Regensburg. Regensburg 1895–1897. pp.viii.143+iv.40. [4000.]

LIST of books, &c., relating to botany and
forestry, including the Cleghorn memorial library, in the library of the museum. Edinburgh
museum of science and art: Edinburgh 1897. pp.
iv.199. [3000.]

[BENJAMIN DAYDON JACKSON], Catalogue of the
library of the Royal botanic gardens, Kew. Royal
gardens, Kew: Bulletin of miscellaneous information (additional series, vol.iii): 1899. pp.viii.ff.790.
[15,000.]

[JOSEPHINE A. CLARK], Catalogue of publications

relating to botany in the library. U.S. Department of agriculture library: Bulletin (no.42): Washington 1902. pp.242. [2750.]

BOTANY. International catalogue of scientific literature (section M): Royal society.

    i. 1902–1903. pp.xiv.378+xiv.626. [5670.]
    ii. 1904. pp.viii.1114. [6339.]
    iii. 1905. pp.viii.909. [5056.]
    iv. 1906. pp.viii.951. [4728.]
    v. 1908. pp.viii.1210. [6219.]
    vi. 1908. pp.viii.843. [6853.]
    vii. 1910. pp.viii.986. [6997.]
    viii. 1910. pp.viii.949. [6517.]
    ix. 1911. pp.viii.859. [6315.]
    x. 1913. pp.viii.840. [7355.]
    xi. 1914. pp.viii.856. [6966.]
    xii. 1915. pp.viii.835. [6730.]
    xiii. 1916. pp.viii.812. [6151.]
    xiv. 1919. pp.viii.753. [5473.]
no more published.

ALICE CARY ATWOOD, Catalogue of the botanical library of John Donnell Smith presented in 1905 to the Smithsonian institution. National museum. Contributions from the United States national herbarium (vol.xii, part i): Washington 1908. pp.iii.94. [1250.]

C. SCHUSTER, Katalog der bibliothek des Botanischen vereins der provinz Brandenburg. Dahlem–Steglitz 1911. pp.viii.192. [3000.]

BIBLIOGRAPHY relating to the floras. Lloyd library: Bibliographical contributions (no.2 &c.): Cincinnati.

Europe in general and the floras of Great Britain. . . . (no.2): 1911. pp.[ii].70. [1000.]

Austria, Bohemia, Poland, Hungary, Belbium, Luxemburg, Netherlands, and Switzerland. . . . (no.3): 1911. pp.[ii].71–132. [600.]

France. . . . (no.4): 1911. pp.[ii].133–186. [750.]

Germany. . . . (no.5): 1912. pp.[ii].187–262. [1000.]

Italy, Spain, Portugal, Greece, european Turkey, Bulgaria, Montenegro, Moldavia, Roumania and Servia. . . . (no.6): 1912. pp.[iii].265–307. [600.]

Arctic regions, Iceland, Scandinavia, Denmark, Norway, Sweden, Russia, Finland, Lapland, russian Poland, and Caucasia. . . . (nos.7–8): 1912. pp.[ii].311–354. [600.]

North America and the West Indies. . . . (no.9): 1913. pp.[ii].355–417. [1000.]

South America and the antarctic regions. . . .

(no.10): 1913. pp.[ii].419–437. [250.]
Asia.... (no.11): 1913. pp.[ii].439–468. [400.]
Oceania. . . . (no.12): 1913. pp.[ii].469–492.
[300.]
Africa. . . . (no.13): 1914. pp.[ii].493–513.
[250.]

CATALOGUE of the books and pamphlets of the
Lloyd library. Lloyd library: Bibliographical contributions (nos.15–32): 1914–1918. pp.[ii].125–800
+[ii].340. [17,500.]

an added title reads Bibliography relating to
botany exclusive of floras.

BOTANICAL abstracts. A monthly serial furnish-
ing abstracts and citations of publications in the
international field of botany in its broadest sense.
Baltimore.

i. 1918–1919. Burton E. Livingston, editor.
pp.xiv.276. [1681.]
ii. 1919. pp.xx.240. [1371.]
iii. 1920. pp.iv.490. [3061.]
iv. 1920. pp.iv.306. [1853.]
v. 1920. pp.vi.357. [2426.]
vi. 1920–1921. pp.vi.329. [2032.]
vii. 1921. pp.vi.371. [2271.]
viii. 1921. pp.vi.349. [2267.]
ix. 1921. pp.vi.286. [1683.]

x. 1921–1922. pp.xxii.333. [2066.]

— Cumulated . . . indexes for volumes 1–10. 1924. pp.[ii].418.

xi. 1922. pp.vii.868. [4895.]

xii. 1923. pp.v.1116. [6737.]

xiii. 1924. pp.v.1242. [8113.]

xiv. 1925. pp.vii.1482. [9686.]

xv. 1926. pp.iv.1262. [9929.]

continued as part of Biological abstracts.

CATALOGUE of early works on botany, agriculture and horticulture. Public library: Cardiff 1919. pp.24. [60.]

A. M. DAVIDSON, Books on botanical subjects bequeathed to the University library by professor [J. W. H.] Trail. [Aberdeen university: Aberdeen 1923]. pp.56. [1250.]

FORTSCHRITTE der botanik. Berlin [&c.].

i. 1931. 1932. Herausgegeben von Fritz von Wettstein. pp.v.264. [1000.]

ii. 1932. 1933. pp.iv.302. [1000.]

iii. 1933. 1934. pp.iv.258. [1000.]

iv. 1934. 1935. pp.iv.326. [1000.]

v. 1935. 1936. pp.iv.346. [1000.]

vi. 1936. 1937. pp.iv.354. [1000.]

vii. 1937. 1938. pp.iv.340. [1000.]

viii. 1938. 1939. pp.iv.346. [1000.]

ix. 1939. 1940. pp.iv.474. [1500.]

x. 1940. 1941. pp.iv.320. [1000.]

xi. 1941. 1944. pp.iv.334. [1000.]

xii. 1942–1948. Herausgegeben von Ernst [Albert] Gäumann ... Otto Renner. 1949. pp.iv.447. [1000.]

xiii. 1949–1950. 1951. pp.iv.388. [1000.]

xiv. 1951. 1953. pp.iv.520. [1500.]

xv. 1952. 1954. pp.iv.549. [1500.]

xvi. 1953. 1954. pp.iv.380. [1000.]

xvii. 1954. 1955. pp.vi.890. [2500.]

xviii. 1955. 1956. pp.vi.385. [1000.]

xix. 1956. 1957. pp.iv.430. [1000.]

xx. 1957. 1958. pp.iv.326. [1000.]

xxi. 1958. 1959. pp.vii.484. [2000.]

xxii. 1959.

xxiii. 1960. 1961. pp.vii.532. [2000.]

xxiv. 1961. 1962. pp.vii.539. [2000.]

xxv. 1962. 1963. pp.vii.619. [2000.]

in progress.

READERS' guide to books on biology, botany and zoology. Library association: County libraries section [no.17]: 1938. pp.[iv].27. [600.]

READERS' guide to botany. Library association: Country libraries section: [Readers' guide (new series, no.11): 1951]. pp.30. [350.]

CATALOGUE of the library at the Chelsea physic garden. 1956. pp.35. [600.]

БОТАНИКА. Каталог (1948–1959). Академия наук СССР: Москва 1959. pp.47. [200.]

GEOGRAPHICAL guide to floras of the world. An annotated list with special reference to useful plants and common plant names. New York &c.

    i. Africa, Australia, north America, south America, and islands of the Atlantic, Pacific, and Indian oceans. By S[idney] F[ay] Blake . . . and Alice C[ary] Atwood. 1963. pp.[iv].336. [3000.]

in progress; also issued as Miscellaneous publication no.401 of the Department of agriculture.

5. *Countries &c.*

[*see also the* Bibliography relating to the floras, *1911–1914, above*].

America, latin, see Tropics

America, north

SERENO WATSON, Bibliographical index to north american botany: or citations of authorities for all the recorded indigenous and naturalized species of the flora of north America. . . . Part I. Polypetalæ. Smithsonian institution: Miscellaneous collections

(no.258): Washington 1878. pp.[v].476. [50,000.]
no more published.

Argentina

F[REDERICO] KURTZ, Essai d'une bibliographie
botanique de l'Argentine. Buenos-Ayres 1900.
pp.91. [650.]

A[LBERTO] CASTELLANOS and R[OMAN] A. PÉREZ-
MOREAU, Contribución a la bibliografía botánica
argentina. I. Buenos Aires 1941 [1942]. pp.555.
[5000.]

no more published.

Asia

V[LADIMIR] L[EONTEVICH] KOMAROV, Библиография к флоре и описанию растиельнойти
дальнего востока. Государственное русской
географическое общество: Записки юожно
уссурийского отдела (vol.ii): Владивосток
1928. pp.278. [1202.]

ELMER D[REW] MERRILL and EGBERT H[AMILTON]
WALKER, A bibliography of eastern asiatic botany.
Harvard university: Arnold arboretum: Jamaica
Plain, Mass. 1938. pp.xlii.719. [35,000.]

Austria

BERICHT über die österreichische literatur der

zoologie, botanik und palaeontologie aus den jahren 1850, 1851, 1852, 1853. Zoologisch-botanischer verein: Wien 1855. pp.vi.376. [1000.]
no more published.

Brazil

ERMINIO MIGLIORATO GARAVINI, Apontamentos e materiaes para um repertorio historico e bio-bibliographico da botanica no Brasil. Primeiro fasciculo. Roma 1913. pp.26. [100.]
no more published?

BIBLIOGRAFIA brasileira de botânica. Instituto brasileiro de bibliografia e documentação: Rio de Janeiro.

    1950–1955. [1957]. pp.153. [854.]
    1956–1958. [1961]. pp.102. [586.]
in progress.

Canada

JACQUES ROUSSEAU, MARCELLE GAUVREAU and CLAIRE MORIN, Bibliographie des travaux botaniques contenus dans les 'Mémoires et comptes rendus de la Société royale du Canada', de 1882 à 1936 inclusivement. Université de Montréal: Institut botanique: Contributions (no.33): Montréal &c. 1939. pp.117. [306.]

ERNEST ROULEAU, Bibliographie des travaux

Botany

concernant la flore canadienne parus dans "Rhodora" de 1899 à 1943 inclusivement. Université de Montréal: Institut botanique: Contributions (no. 54): Montréal 1944. pp.3–367.

Ceylon

T[OM] PETCH, Bibliography of books and papers relating to agriculture and botany to the end of the year 1915. Peradeniya manuals (vol.iii): Colombo 1925. pp.[iii].256. [5059.]

China

LIOU-HO and [J. A.] CLAUDIUS ROUX, Aperçu bibliographique sur les anciens traités chinois de botanique, d'agriculture, de sériculture et de fungiculture. Lyon 1927. pp.39. [75.]

Czechoslovakia

J[ÁN] FUTÁK and K[AROL] DOMIN, Bibliografia k flóre ČSR do r. 1952. Slovenská adadémie vied: Sekcia biologických a lekárskych vied: Bratislava 1960. pp.884. [20,000.]

JINDŘICH HOUFEK and VLADIMIR SKALICKÝ, Botanická bibliografie jižních čech. Sborník Jihočeského muzea: České Budějovice 1963. pp.127.

Botany

Denmark

CARL F[REDERIK] [ALBERT] CHRISTENSEN, Den danske botaniske litteratur, 1880–1911. København 1913. pp.xxiii.279. [2500.]

—— 1912–1939. 1940. pp.[vii].350. [6500.]

Estonia

BOTAANILINE kirjandus 1945–1955. Eesti NSV teaduste akadeemia: Zoologia ja botaanika instituut: Tartu 1958. pp.36. [278.]

France

A[RTHUR] L[OUIS] LETACQ, Notice sur quelques botanistes ornais et essai sur la bibliographie botanique du département de l'Orne. Caen 1889. pp.66. [180.]

C[HARLES] BRUYANT, Bibliographie raisonnée de la faune et de la flore de l'Auvergne. 1894. pp.[ii]. 91. [75.]

A. TRONCHET, Aperçu historique et bibliographique sur la floristique et la phytosociologie en Franche-Comté et régions limitrophes. Besançon [printed] [1949]. pp.13. [75.]

Germany

AUGUST [ALBERT HEINRICH] SCHULZ, Die floristi-sche litteratur für Nordthüringen, den Harz und den provinzialsächsischen wie anhaltischen teil an der norddeutschen tiefebene. Halle a. S. 1888. pp.90. [800.]

—— Zweite ... auflage. 1891. pp.90.22. [1000.]

FERDINAND PAX, Bibliographie der schlesischen botanik. Historische kommission für Schlesien: Schlesische bibliographie (vol.iv): Breslau 1929. pp.[viii].167. [2527.]

Great Britain

JAMES BRITTEN and G[EORGE] S[IMONDS] BOULGER, A bibliographical index of british and irish botanists. 1893. pp.xv.188. [5000.]

—— First supplement (1893–1897). 1899. pp. 189–222. [750.]

—— Second supplement (1898–1902). 1905. pp.19. [400.]

BIBLIOGRAPHY of Worcestershire. Part III. Works relating to the botany of Worcestershire. Compiled by John Humphreys. Worcestershire Historical Society: Oxford [printed] 1907. pp.[iii]. 217–252. [350.]

JOHN SMART, Bibliography of key works for the identification of the british fauna and flora. Association for the study of systematics in relation to

general biology: Publication (no.1): 1942. pp.viii.
105. [1000.]

—— Second edition. Edited by John Smart
and George Taylor. 1953. pp.xi.126. [1250.]

N[ORMAN] DOUGLAS SIMPSON, A bibliographical
index of the british flora, including floras, herbals,
periodicals, societies and references relating to the
identification, distribution and occurrence of pha-
nerogams, vascular cryptogams and cherophytes
in the British Isles. Bournemouth [printed] 1960.
pp.xix.429. [35,000.]

privately printed.

India

LIST of publications on the botany of indian
crops.

[i].   –1927. Imperial institute of agricul-
          tural research: Bulletin (no.202): Pusa 1930.
ii. 1928–1932. By R. D. Bose [Rāhhāldās
          Vasu]. Imperial council of agricultural
          research: Miscellaneous bulletin (no.12):
          1936. pp.[v].198. [3000.]

Indochina

A. PÉTELOT, Analyse des travaux de zoologie et
de botanique concernant l'Indochine publiés en

1929. Direction générale de l'instruction publique: Hanoï 1930. pp.[i]22. [60.]

Italy

P[IETRO] A[NDREA] SACCARDO, La botanica in Italia. . . . I. Repertorio biografico e bibliografico dei botanici italiani, aggiuntivi gli stranieri che trattarono della flora italiana. II. Indici dei floristi d'Italia disposti secondo le regioni esplorate. III. Cenni storici e bibliografici degli orti botanici publici e privati. IV. Quadro cronologico dei principali fatti botanici ne' quali gli italiani furono precursori. Reale istituto veneto di scienze, lettere ed arti (vol.xxv, no.4): Venezia 1895. pp.236. [5000.]

BULLETTINO bibliografico della botanica italiana. Società botanica italiana: Firenze.

anno

    i. 1904. Redatto per cura del dott. G. B. Traverso. pp.viii.80. [666.]

    ii. 1905. pp.76. [622.]

    iii. 1906. pp.64. [506.]

    iv. 1907. pp.72. [594.]

    v. 1908. pp.79. [669.]

    vi. 1909. pp.64. [558.]

    vii. 1910. pp.120. [1055.]

    viii. 1911. pp.121–194. [645.]

ix. 1912. pp.195–278. [751.]
x. 1913. pp.279–356. [717.]
xi. 1914. pp.75. [699.]
xii. 1915. pp.77–162. [813.]
xiii. 1916. pp.163–230. [634.]
volume
iv. 1917–1923. Redatto per cura del dott. Cesare Sibilia. pp.x.336. [3613.]
v, fasc. I. 1924. pp.[ii].154. [652.]
v, fasc. II. 1925. pp.[ii].60. [773.]
[v, fasc. III.] 1926. pp.75. [623.]
vi. 1927–1929. Redatto per cura della dott. Albina Messeri. pp.[ii].152. [1488.]
vii. 1930–1932. pp.[ii].55.53.69. [1773.]
viii. 1933–1935. Redatto per cura della prof. A. Messeri . . . e del dott. Rodolfo Pichi Sermolli. pp.[ii].51.47.56. [1552.]
ix. 1936–1938. pp.[iii].71.58.63. [1822.]

AUGUSTO BÉGUINOT, La botanica. Guide "ICS": Roma 1920. pp.[iv].116. [1000.]

Japan

HARVEY HARRIS BARTLETT and HIDE SHOHARA, Japanese botany during the period of wood-block printing. . . . An exhibition of japanese books & manuscripts, mostly botanical, held at the Clements library [Ann Arbor]. Los Angeles 1961. pp.vi.271. [112.]

Korea

ALICE MARIA GOODE, An annotated bibliography of the flora of Korea. Catholic university of America: Washington 1955. ff.[v].65.3. [305.]*

Melanesia, see New Guinea

Mexico

MANUEL DE OLAGUÍBEL, Memoria para una bibliografía científica de México en el siglo XIX. México 1889. pp.99. [500.]
only the first section, on botany, was published.

NICOLÁS LEÓN, Bibliotheca botánica-mexicana. Catálogo bibliográfico, biográfico y crítico de autores y escritores referentes a vegetales de México y sus aplicaciones, desde la conquista hasta el presente. México 1895. pp.372. [1000.]

Micronesia

MARIE HÉLÈNE SACHET and F[RANCIS] RAYMOND FOSBERG, Island bibliographies, Micronesian botany, land environment and ecology of coral atolls, vegetation of the tropical Pacific islands. Compiled under the auspices of the Pacific science

Botany

board. National academy of sciences: National research council: Publication (no.335): Washington 1955. pp.[iii].v.577. [7500.]*

New Guinea

SELECTED list of references to publications in the Science library on the fauna and flora of New Guinea and the Melanesian Islands. Science library: Bibliographical series (no.705): 1951. ff.2. [23.]*

Pacific

ELMER D[REW] MERRILL, A botanical bibliography of the islands of the Pacific. United States national herbarium: Contributions (vol.xxx, part I): Washington 1947. pp.v.404. [3850.]

Philippines

ELMER D[REW] MERRILL, A discussion and bibliography of Philippine flowering plants. Bureau of science: Popular bulletin (no.2): Manila 1926. pp.239. [1750.]

Poland

DEZYDERY SZYMKIEWICZ, Bibliografja flory polskiej. Polska akademia umiejętności: Prace monograficzne Komisji fizjograficznej (vol.ii): Krakowie 1925. pp.158. [2035.]

Botany

Polynesia

E[LMER] D[REW] MERRILL, Bibliography of polynesian botany. Bernice P. Bishop museum: Bulletin (no.13): Honolulu 1924. pp.68. [1292.]

— — [another edition]. . . . (no.144): 1937. pp. 194. [2600.]

Russia

E. R. TRAUTVETTER, Grundriss einer geschichte der botanik in bezug auf Russland. St. Petersburg 1837. pp.[ii].145. [750.]

S. M. VISLOUKH [*and others*], Обзоръ ботанико-географической литературы по флорѣ Россіи за 1906 годъ. С.-Петербургъ 1908. pp.69. [300.]

D. I. LITVINOV, Библіографія флоры Сибири. С.-Петербургъ 1909. pp.ix.460.

M. G. MIKHAILOVA, Флора та рослинність УРСР. Бібліографія. Академія Наук УРСР: Інститут Ботаніки: Київ 1938. pp.64. [1706.]

S[ERGEI] YU[LEVICH] LIPSHITS, Русские ботаники. Биографо-библиографический словарь. Московское общество испытателей природы [&c.] Москва.

 i. А-[Бушинский]. 1947. pp.xi.336. [10,000.]
 ii. Быков-Горленко 1947. pp.336. [10,000.]

iii. Горницкий-И.1950. pp.vii.488. [15,000.]
iv. К-Кюз. 1952. pp.644. [25,000.]
in progress.

D. V. LEBEDEV, Введение в ботаническую литературу СССР. Академия наук СССР: Ботанический институт им. В. Л. Комарова: Москва &c. 1956. pp.383. [3000.]

M. D. BARUISHEVA [*and others*], Обощевод-ство и картофелеводство на Крайнем севере. Библиографический указатель, 1932–1957 гг. Центральная научная сельскохозяйствен-ная библиотека [&c.]: Ленинград 1959. pp.51.

Spain and Portugal

MIGUEL COLMEIRO, La botánica y los botánicos de la península hispano-lusitana. Estudios biblio-gráficos y biográficos. Madrid 1858. pp.xi.216. [932.]

Sweden

JOHANNES EM[ANUEL] WIKSTRÖM, Conspectus litteraturæ botanicæ in Suecia ab antiquissimis temporibus usque ad finem anni 1831. Holmiae 1831. pp.[ii].xlix.342. [1500.]

TH[ORGNY] O[SSIAN] B[OLIVAR] N[APOLEON] KROK, Bibliotheca botanica suecana ab antiquissimis temporibus ad finem anni MCMXVIII. Uppsala &c. 1925. pp.xvi.799. [15,000.]

Switzerland

ED. FISCHER, Flora helvetica, 1530–1900. Centralkommission für schweizerische landeskunde: Bibliographie der schweizerischen landeskunde (section IV.5): Bern 1901. pp.xviii.241. [4000.]

—— Nachträge. 1922. pp.ix.40. [600.]

JOHN BRIQUET, Biographies des botanistes à Genève de 1500 à 1931. Société suisse: Bulletin (vol.502): Genève 1940. pp.x.494. [4000.]

Tropics

J[OHN] C[HRISTOPHER] WILLIS, Literature of tropical economic botany & agriculture. II. 1904–1908 to 1907–1910. [Colombo] 1911. pp.45. [1500.]

HELEN V[IRGINIA] BARNES and JESSIE M[AY] ALLEN, A bibliography of plant pathology in the tropics and in latin America. Department of agriculture: Bibliographical bulletin (no.14): Washington 1951. pp.vi.78. [2395.]

TROPICAL vegetation. List of references, India, Burma & Ceylon 1948–1954. Unesco: South Asia

science co-operation office: [New Delhi 1956].
pp.[iii].12. [300.]*

Tunisia

LISTE des publications disponibles. Service botanique et agronomique de Tunisie: Ariana [1951].
pp.24. [350.]

— Supplément. [1954]. pp.4. [50.]

United States

PAUL C[ARPENTER] STANDLEY, The type localities
of plants first described from New Mexico: a
bibliography of new mexican botany. Smithsonian institution: United States national museum:
Contributions from the U.S. national herbarium
(vol.xiii, part 6): Washington 1910. pp.143–246.
xiv. [340.]

A LIST of references on the flora of the south-
eastern United States. Library of Congress: Washington 1932. ff.8. [77.]*

HOMER D[OLIVER] HOUSE, Bibliography of botany of New York state, 1751–1940. New York
state museum: Bulletin (nos.328–329): Albany
1941–1942. pp.174+177–233. [3000.]

HAROLD T. PINKETT, Preliminary inventory of
the records of the Bureau of plant industry, soils
and agricultural engineering. National archives:

Preliminary inventories (no.66): Washington 1954. pp.v.49. [very large number.]★

s[IDNEY] F[AY] BLAKE, Guide to popular floras of the United States and Alaska. An annotated, selected list of nontechnical works for the identification of flowers, ferns and trees. Department of agriculture: Bibliographical bulletin (no.23): Washington 1954. pp.[ii].56. [300.]

H[ENRY] C[AMPBELL] GREENE and J[OHN] T[HOMAS] CURTIS, A bibliography of Wisconsin vegetation. Public museum: Publications in botany (no.1): Milwaukee 1955. pp.84. [750.]

EARL L[EMLEY] CORE, WILLIAM H. GILLESPIE and BETTY J. GILLESPIE, Bibliography of West Virginia plant life. SL bibliography series: New York [1962]. pp.[v].46. [900.]★

Wales

IRENE D. REES, Bibliography of articles, reports and scientific papers by members of staff 1919–1955. University college of Wales: Welsh plant breeding station: Aberystwyth 1956. pp.[ii].98. [782.]

6. *Classification*

[ALFRED BARTON RENDLE], Books and portraits illustrating the history of plant classification exhibited in the Department of botany. British museum (Natural history): Special guides (no.2): 1906. pp.19. [40.]

[—] — Second edition. 1909. pp.20. [50.]

7. *Distribution*

A[UGUST HEINRICH RUDOLF] GRISEBACH, Bericht über die leistungen in der pflanzengeographie [1845, 1847: und systematischen botanik; 1848–1853: geographischen und systematischen botanik] während des jahres. Berlin.

1843. 1845. pp.[ii].78. [100.]
1844. 1846. pp.[ii].88. [100.]
1845. 1847. pp.[ii].78. [100.]
1846. 1849. pp.[ii].64. [100.]
1847. 1850. pp.[ii].94. [100.]
1848. 1851. pp.[iii].107. [200.]
1849. 1851. pp.101. [200.]
1850. 1853. pp.[ii].120. [200.]
1851. 1854. pp.[ii].122. [200.]
1852. 1855. pp.[ii].126. [250.]
1853. 1856. pp.[ii].98. [250.]

FRANK EDWIN EGLER, A cartographic guide to selected regional vegetation literature, where plant communities have been described. Université de

Montréal: Institut botanique: Sarracenia (no.1): Montréal 1959. ff.50.*

500 copies reproduced from typewriting.

8. *Economics*

BENJAMIN DAYDON JACKSON, Vegetable technology; a contribution towards a bibliography of economic botany. . . . Founded upon the collections of George James Symons. Index society: Publications (vol.xi): 1882. pp.xii.356. [3000.]

J. E. ROCKWELL, Index to papers relating to plant-industry subjects in the yearbooks of the United States department of agriculture. Department of agriculture: Bureau of plant industry: Circular (no.17): Washington 1908. pp.55. [2000.]

JESSIE M. ALLEN, Check list of publications issued by the Bureau of plant industry, United States department of agriculture, 1901–1920, and by the divisions and offices which combined to form this Bureau, 1862–1901. Department of agriculture library: Bibliographical contributions (no.3): Washington 1921. pp.127. [4000.]*

9. *Genetics*

FRIEDRICH OEHLKERS, Erblichkeitsforschung an pflanzen. Ein abriss ihrer entwicklung in den letzten 15 jahren. Wissenschaftliche forschungsberichte: Naturwissenschaftliche reihe (vol.xviii): Dresden &c. 1927. pp.viii.204. [400.]

HAJIME MATSUURA, A bibliographical monograph on plant genetics (genic analysis), 1900–1925. Imperial university: Contribution to cytology and genetics (no.82): Tokyo 1929. pp.[ii]. xii.499. [1341.]
— — Second edition. Hokkaido imperial university: Sapporo 1933. pp.[ii].xx.789. [2077.]

PLANT breeding abstracts. Imperial bureau of plant [breeding and] genetics: School of agriculture: Cambridge.
   i. 1930–1931. pp.[v].24.26.52.40. [606.]
   ii. 1931–1932. pp.[vi].220. [726.]
   — Indexes to volumes I and II. 1932. pp.35.
   iii. 1932–1933. pp.[viii].224. [759.]
   — Indexes to volumes I to III. 1933. pp.44.
   iv. 1933–1934. pp.[vii].360. [1127].
   v. 1934–1935. pp.[viii].380. [1159.]
   — Indexes to volumes I to v. 1935. pp.71.
   vi. 1935–1936. pp.[ix].464+47. [1449.]
   vii. 1936–1937. pp.[ix].449. [1398.]
   — Indexes to volumes VI and VII. 1937. pp.66.
   viii. 1937–1938. pp.[x].509. [1662.]
   ix. 1939. pp.[x].492. [1662.]

x. 1940. pp.[x].325+34. [1163.]
xi. 1941. pp.[ii].338+31. [1145.]
xii. 1942. pp.[ii].296+33. [1236.]
xiii. 1943. pp.[ii].384+30. [1432.]
xiv. 1944. pp.[ii].344+24. [1347.]
xv. 1945. pp.[ii].381+25. [1539.]
xvi. 1946. pp.[ii].494+29. [1956.]
xvii. 1947. pp.[ii].504+29. [1823.]
xviii. 1948. pp.[ii].838+35. [2577.]
xix. 1949. pp.[ii].942+38. [2998.]
xx. 1950. pp.[ii].846+37. [2716.]
xxi. 1951. pp.[ii].1007+40. [3160.]
xxii. 1952. pp.[ii].640+39. [3096.]
xxiii. 1953. pp.[ii].679+42. [3127.]
xxiv. 1954. pp.662+   . [3506.]
xxv. 1955. pp.[ii].xxii.646.50. [3546.]
xxvi. 1956. pp.[ii].720.60. [3946.]
xxvii. 1957. pp.[ii].770.80. [4624.]
xxviii. 1958. pp.[ii].972. [4947.]
xxix. 1959. pp.[ii].1012. [4720.]
xxx. 1960. pp.[ii].962. [4550.]
xxxi. 1961. pp.[ii].1138. [5630.]
xxxii. 1962. pp.[ii].1116. [5723.]
xxxiii. 1963. pp.[ii].   . [5368.]

— Supplement 1. Summary of reports received from countries exclusive of the

British Empire, 1928–1931. 1933. pp.44.
[500.]
— Supplement II. Summary of reports received from stations in the British Empire,
1932–1935. 1936. pp.64. [600.]
in progress.

MARJORIE F[LEMING] WARNER, MARTHA A[LVORD]
SHERMAN and ESTHER M[ARIE] COLVIN, A bibliography of plant genetics. Department of agriculture: Miscellaneous publication (no.164): Washington 1934. pp.552. [10,156.]

FULL translations filed at the Imperial bureau of
plant genetics. Cambridge 1932. ff.[i].5. [25.]★
— [another edition]. Abridged and full translations [&c.]. 1934. ff.[i].8. [35.]★
— [another edition]. 1939. ff.[i].6. [23.]★

TRANSLATIONS filed in the library and available
on loan. Imperial bureau of plant genetics (for
crops other than herbage): Aberystwyth 1945 &c.★

H[ANS] STUBBE, Genetisch-pflanzenzüchterische
bibliographie, 1939–1946 (1947). Der Züchter
(sonderheft): Berlin 1949. pp.287. [11,000.]

10. *Iconography*

CLAUS NISSEN, Botanische prachtwerke. Die

blütezeit der pflanzenillustration von 1740 bis
1840. Wien 1933. pp.45. [353.]

[RUTH SHEPARD GRANNISS], Plant illustration before 1850. A catalogue of an exhibition of books,
drawings and prints held by the Garden club of
America and the Grolier club. Grolier club: New
York 1941. pp.iii–xiii.33. [134.]

[MONCURE BIDDLE], A Christmas letter. Some
flower books and their makers. Philadelphia 1945.
pp.49. [50.]

CLAUS NISSEN, Die botanische buchillustration,
ihre geschichte und bibliographie. Stuttgart 1951
[–1952]. pp.vii.264+[vii].324. [2387.]

BOTANICAL books, prints & drawings from the
collection of mrs. Roy Arthur Hunt. Carnegie
institute: Department of fine arts: Pittsburgh 1952.
pp.xvi.63.[xv]. [86.]

SACHEVERELL SITWELL and WILFRID BLUNT, Great
flower books 1700–1900. A bibliographical record
of two centuries of finely-illustrated flower books.
. . . The bibliography edited by Patrick M[illing-
ton] Synge. 1956. pp.x.94. [750.]

11. *Medicinal plants*

MARIA CHMIELIŃSKA, Polska bibliografia zie-

larstwa, za okres od początku XVI wieku do roku
1940. Warszawa 1954. pp.xxxii.516. [5193.]

L. A. UTKIN, A. F. GAMMERMAN and V. A. NEVSKY,
Библиография по лекарственным растениям.
Указатель отечественной литературы: ру-
кописи XVII–XIX вв., печатные издания 1732–
1954 гг. Академия наук СССР: Ботанический
институт им. В. Л. Комарова: Москва &c.
1957. pp.727. [9000.]

LIBRARY of medicinal plants collected by Henry
G. de Laszlo. Cambridge 1958. pp.[56]. [1200.]*

12. *Paleontology*

DIE PALAEOBOTANISCHE literatur. Bibliographi-
sche übersicht über die arbeiten aus dem gebiete
der palaeobotanik. Jena.

    i. 1908. Herausgegeben von W[illem J[oseph]
        Jongmans. pp.iv.218. [6000.]
    ii. 1909. pp.[iii].418. [12,000.]
    iii. 1910–1911. pp.[ii].570. [16,000.]

FOSSILIUM catalogus. II: Plantae. Editus a W[il-
lem Joseph] Jongmans. Berlin [part &c.:
's-Gravenhage] 1913 &c.

REPORT of the Committee on paleobotany . . .
with bibliography of paleobotany in north and

south America. National research council: Division of geology and geography: Committee on paleobotany: Washington 1951. pp.[v].41. [250.]★

PALYNOLOGIE. Bibliographie. Museum national d'histoire naturelle: Centre de documentation des sciences de la terre et section de palynologie.

    i. 1956. pp.116. [575.]
    ii. 1958. pp.116. [722.]
    iii. 1959. pp.132. [523.]
no more published.

FRANTIŠEK NĚMEJC, Úvod do starší paleobotanické literatury. Ceskoslovenská akademie věd: Sekce geologicko-geografická: Praha1960. pp.140. [2000.]

M[ARK] I[LICH] NEISHTADT, Палинология в СССР (1952–1957 гг.). Академия наук СССР: Институт географии: Москва 1960. pp.272. [1632.]

13. *Pathology*

[*see also* **Entomology: Agriculture**
and **Mycology**]

i. *Periodicals*

A LIST of periodicals dealing wholly or mainly with phytopathology. Science library: Bibliographical series (no.239): 1936. ff.4. [87.]★

Botany

ii. *General*

CATALOGUS van de bibliotheek. Phytopathologisch laboratorium "Willie Commelin Scholten": Amsterdam 1909. pp.39. [600.]

CHARLES C. REES and WALLACE MACFARLANE, A bibliography of recent literature concerning plant-disease prevention.... And a bibliography of non parasitic diseases of plants by Cyrus W. Lantz. University of Illinois: Agricultural experiment station: Circular (no.183): 1915. pp.[ii].111. [1250.]

A CHECK list of the publications of the Department of agriculture on the subject of plant pathology, 1837–1918. Department of agriculture: Bibliographical contributions (no.1): [Washington] 1919. pp.[ii].38. [600.]*

— [another edition]. Jessie M. Allen, Author and subject index to the publications [&c.]. . . . (no.10): 1928. pp.[iii].xiii.251. [2750.]*

[EUNICE ROCKWOOD OBERLY and JESSIE MAY ALLEN], Check list of the state agricultural experiment stations on the subject of plant pathology. United States department of agriculture library: Bibliographical contributions (no.2): Washington 1922. pp.[iv].1–142.119–179. [5000.]*

BREEDING varieties resistant to disease. Imperial bureau of plant genetics: School of agriculture: Cambridge [1930]. ff.[i].ii.43. [300.]*

— Breeding resistant varieties, 1930–1933. List of titles supplementary to the bibliography on breeding varieties. [1934]. pp.[ii].32. [300.]

G[EOFFREY] C[LOUGH] AINSWORTH, The plant diseases of Great Britain. A bibliography. 1937. pp.xii.273. [1750.]

CARL J[OHN] DRAKE, Bibliography of the state plant pest laws, quarantines, regulations and administrative rulings of the United States of America. National plant board: [Ames, Ia. 1942]. ff.[i]. 60. [500.]*

R. MARTINEZ CROVETTO and B. G. PICCININI, Bibliografía argentina sobre malezas. Instituto de botanica: Publicación tecnica (n.s., no.17) Buenos Aires 1948. pp.91. [600.]

PLANT diseases and plant pests: a select list of U.S. government publications. State library: Smithsonian bibliography (no.28 = Agriculture series, no.3): Pretoria 1951. ff.[ii].19. [257.]*

SOVIET publications in plant pathology and closely related fields. Titles published from 1925 through early 1953 and available in the library of the U.S. Department of agriculture. [Washington 1953]. pp.iv.336. [2563.]*

A[LLEN] W. SCHOLL and DANA RAY CARTWRIGHT, Bibliography of synthetic plant-growth substances. [Huntington, W. Va.] 1955. pp.vii.532. [2743.]*

D[ENISE] SCHEIDECKER, Diagnostic foliaire. Éléments de documentation. Office de la recherche scientifique et technique outre-mer: Centre de phytophysiologie: 1955. ff.[ii].44. [350.]*

HAROLD T. PINKETT, Preliminary inventory of the records of the Bureau of entomology and plant quarantine. National archives: Preliminary inventories (no.94): Washington 1956. pp.v.24. [large number.]*

14. *Physiology and chemistry*

F[RANZ] J[ULIUS] F[ERDINAND] MEYEN, Jahresbericht über die resultate der arbeiten im felde der physiologischen botanik. Berlin.

 1837. 1838. pp.vi.186. [100.]
 1838. 1839. pp.vi.154. [125.]
 1839. 1840. pp.viii.184. [175.]
   [continued as:]
Jahresbericht über die arbeiten für physiologische botanik. Berlin.

1840. By H[einrich] F[riedrich] Link. 1842.
pp.[ii].102. [60.]
1841. 1843. pp.[ii].80. [50.]
1842–1843. 1844. pp.[ii].133. [150.]
1844–1845. 1846. pp.[ii].113. [125.]
[*continued as:*]
Jahresbericht über die leistungen im gebiete der physiologischen botanik.

1846. By Julius Münster. 1849. pp.[ii].128.
[300.]

IVOR GRIFFITH, A half century of research in plant chemistry. A chronological record of the scientific contributions of Frederick Belding Power. [Philadelphia 1924]. pp.16. [127.]

REFERENCES on the carbon nitrogen ratio in plants. Science library: Bibliographical series (no.27): [1931]. ff.[i].3. [44.]★

H[ANS] NIKLAS [*and others*], Literatursammlung aus dem gesamtgebiet der agrikulturchemie. Band III. Pflanzenernährung. Bodenuntersuchungsstelle: Weihenstephan 1934. pp.xlv.1114. [25,000.]

L[UTHER] G[EORGE] WILLIS, Bibliography of references to the literature on the minor elements and their relation to the science of plant nutrition.

Chilean nitrate educational bureau: [New York] 1935. pp.455. [2700.]

—— Third edition. [1939]. pp.[iii].488. [4628.]

[—] — Fourth edition. Bibliography of the literature on the minor elements and their relation to plant and animal nutrition. Volume I. [Edited by Barbara Johnson]. 1948. pp.[iii].1037. [10,000.]

THE PHYSIOLOGICAL action of weedkillers and other plant poisons. Science library: Bibliographical series (no.235): 1936. ff.7. [118.]*

THE EFFECT of chromium compounds on plants and their occurrence in plants. Science library: Bibliographical series (no.424): 1938. ff.3. [41.]*

CATHERINE M[URER] SCHMIDT and DOROTHY H[ARDING] JAMESON, Bibliography of literature on analyses of leaf and other plant tissues, with special reference to content of mineral nutrients, 1935 through 1940. American potash institute: Special bibliography (no.2): Washington 1941. pp.[iv].156.xv. [669.]*

BIBLIOGRAPHY of the literature on sodium and iodine in relation to plant and animal nutrition, Chilean nitrate educational bureau: New York.

i. 1948. pp.[ii].123. [1000.]
no more published.

IODINE and plant life. Annotated bibliography 1813–1949, with review of the literature. Chilean iodine educational bureau: 1950. pp.ix.114. [818.]

N. S. GELMAN and G. D. ZENKEVICH, Биохимия растений. Библиографический указатель отечественной литературы 1738–1952 гг. Академия наук СССР: Отделение биологических [&c.]: Материалы по истории биологических наук в СССР: Москва 1956. pp.395. [8500.]

[MARGARET L. BUCH], A bibliography of organic acids in higher plants. Department of agriculture: Agricultural research service (no.73–18): Washington 1957. pp.136. [750.]

—— [another edition]. Department of agriculture: Agricultural research service: Agricultural handbook (no.164): [1960]. pp.100. [823.]

TOXICITY to plants of arsenic in soil. Science library: Bibliographical series (no.753): 1957. ff.2. [50.]*

MARTIN JACOBSON, Insecticides from plants. A review of the literature, 1941–1952. Department of agriculture: Agriculture handbook (no.154): [Washington 1958]. pp.[iv].299. [2500.]

SOME references to the use of calcium cyanamide

as a weed killer (1932–1958). Science library: Bibliographical series (no.765): [1958]. pp.4. [100.]*

15. *Miscellaneous*

[JACQUES CAMBESSÈDES], Liste des travaux de botanique publiés par m. J. Cambecèdes. [1830]. pp.3. [13.]

[P. É. S. DUCHARTRE], Travaux relatifs à la botanique, par m. P. Duchartre. [1853]. pp.8. [25.]
—— [another edition]. [1861]. pp.[ii].42. [70.]

[E. GERMAIN DE SAINT PIERRE], Notice sur les mémoires et les ouvrages de botanique. 1855. pp.20. [51.]

D'ARCY W[ENTWORTH] THOMPSON, A catalogue of books and papers relating to the fertilisation of flowers. 1883. pp.[ii].36. [814.]

[EDWARD WILLIAM BADGER], A sketch of the botanical work of James E. Bagnall. Birmingham [printed] 1897. pp.16. [13.]

A RECORD of the doctors in botany of the university of Chicago, 1897–1916. Chicago 1916. ff. vii.82. [500.]

A[RON] JUNGELSON, Bibliographie de l'action du cuivre sur les végétaux. Institut agricole de l'université: Bulletin bibliographique de botanique agricole (no.i): Nancy 1917. pp.xvi.61. [1308.]

LIST of references on botanical gardens. Library of Congress: Washington 1917. ff.6. [60.]★

LIST of references on fresh-water fauna & flora (supplementary to bibliography in "The life of inland waters" by Needham & Lloyd, 1916). Library of Congress: Washington 1917. ff.3. [28.]★

BIBLIOGRAPHY on the effect of electricity on plants. Science library: Bibliographical series (no.24): [1931]. ff.[i].3.2. [79.]★

BIBLIOGRAPHY of the effect of artificial light on the growth of plants. Science library: Bibliographical series (no.49): [1932]. pp.[i].3. [50.]★
— [another edition]. Effect of artificial light on plants. . . . (no.315): 1937. ff.9. [161.]★

THE INFLUENCE of the moon on plants. Science library: Bibliographical series (no.73): [1933]. single sheet. [13.]★

THE EFFECT of centrifugal force on plants. Science library: Bibliographical series (no.79): [1933]. ff.[i].2. [21.]★

HORACE S. DEAN, Cumulative index to service and regulatory announcements, nos. 1 to 117 in-

clusive, 1914–1933. Bureau of plant quarantine:
Washington 1934. pp.ii.68. [4000.]

BIBLIOGRAPHIA phytosociologica. Hannovera.
i–ii. Germania, Auctore R. Tüxen.... Neder-
landia, auctore W. C. De Leeuw. 1935.
pp.64. [900.]
iii. Regio mediterranea. Auctore: J. Braun-
Blanquet. 1936. pp.20. [350.]
iv–vi. Germania. Pars II. Auctoribus R.
Tüxen [and others]. [Sudentenländer . . .
von F. Pohl; Slowakei und ehem. Karpa-
then-Russland . . . von F. Pohl]. 1943.
pp.134. [1500.]

RODNEY BEECHER HARVEY, An annotated bibliography of the low temperature relations of plants.
. . . Revised [second] edition. Minneapolis, Minn.
1936. ff.[i].ii.240. [3750.]★

BIBLIOGRAPHY on cold resistance in plants. Imperial bureau of plant breeding and genetics.
School of agriculture: Cambridge [1939]. pp.[ii].
22. [371.]

LEGENDS and anecdotes of flowers: a list of
references prior to 1892. Library of Congress:
Washington 1948. ff.4. [38.]★

[V. F. VERZILOV and D. V. ISAKOVA], Владимир

Botany

Леонтьевич Комаров. Акалемия наук Союза ССР: Материалы к биобиблиографии ученых СССР: Серия биологических наук: Ботаника (vol.1): Москва 1946. pp.100. [1000.]

C. D. DARLINGTON and collaborators. Bibliography. Botany school: Oxford [1958]. pp.32. [400.]

GEORGES PLAISANCE, Les formations végétales et paysages ruraux. Lexique et guide bibliographique. 1959. pp.424. [3500.]

Conservation.

SELECT list of references on conservation of human life in the United States. Library of Congress: Washington 1912. ff.6. [55.]*

LIST of National park publications. Department of the Interior: Office of the secretary: [Washington 1912]. pp.27. [600.]

HERMANN H[ENRY] B[ERNARD] MEYER, Select list of references on the conservation of natural resources in the United States. Library of Congress: Washington 1912. pp.110. [600.]
— — [supplement]. List of references on conservation of human life. 1917. ff.4. [41.]*

LIST of references on the conservation of the natural resources of Europe. Library of Congress: Washington 1914. ff.5. [40.]*

BRIEF list of references on the conservation and preservation of scenery, historic monuments, etc. Library of Congress: Washington 1916. ff.3. [25.]*

BRIEF list of references on cut-over lands, reclamation and reforestation. Library of Congress: Washington 1923. ff.4. [46.]*

LIST of engineering articles. No.8. [Partial list of articles in technical and other periodicals on the Bureau of reclamation]. Bureau of reclamation: Washington 1934. pp.[ii].67. [1500.]

NATURSCHUTZ. Ein besprechendes bücherverzeichnis. Stadtbücherei: Charlottenburg 1935. pp.24. [75.]

MARIAN R. KIEKENAPP, Conservation in Minnesota. University of Minnesota: Bibliographical projects (no.3): [Minneapolis] 1936. ff.[iii].18. [250.]★

ANN DUNCAN BROWN, A list of references on the United States conservation corps. Compiled ... under the direction of Florence S[elma] Hellman. Library of Congress: [Washington] 1936. pp.26. [365.]★
— Supplement. 1937. pp.13. [155.]★
— Supplement. 1939. pp.14. [159.]★

A BIBLIOGRAPHY of national parks and monuments west of the Mississippi. Department of the interior: National park service: [s.l.] 1941 &c.

ETTA G[REENBLATT] RIGERS, Publications and visual information on soil conservation. Department of agriculture: Miscellaneous publication (no.446): [Washington] 1944. pp.20. [400.]

Conservation

EVANGELINE THURBER, Materials in the National archives relating to the work of the Civilian conservation corps. National archives: Reference information circular (no.30): [Washington] 1944. pp.6. [large number.]

PREVENTION of deterioration advance list. National research council: Prevention of deterioration center: Washington 1948 &c.★
in progress.

NATIONAL parks. A bibliography on national parks, footpaths and access, and wild life conservation. House of commons: Library: Bibliography (no.62): [1949]. pp.22. [160.]★

BULLETINS, books and visual aids subjects relating to the conservation of natural resources. Department of conservation: [Lansing] 1951 &c.
in progress.

MURIEL BEUSCHLEIN, Free and inexpressive materials for conservation education. [Chicago 1952]. pp.15. [1000.]

NEIL HOTCHKISS, Wildlife abstracts 1935–51. An annotated bibliography of the publications abstracted in the Wildlife review, nos.1–66. Department of the interior: Fish and wildlife service: Washington 1954. pp.[iv].435. [10,000.]★

c]ORNELIS] H[ENDRIK] EDELMAN and B. E. P. EEUWENS, Bibliography on land and water utilization and conservation in Europe. Food and agriculture organization of the United Nations: Rome 1955. pp.[viii].347. [3500.]★

EDWARD E. HILL, Preliminary inventory of the records of the Bureau of reclamation. National archives: Preliminary inventories (no.109): Washington 1958. pp.v.27. [very large number.]

SELECTED references in conservation education for Texas teachers and pupils. Advisory committee on conservation education: [*s.l.* 1958]. pp.iv.58. [500.]★

BIBLIOGRAFIE ochrany přirody v Československu. Státní ustav památkově péče a ochrany přírody: Praha.

    1958. Sestavil Jiří Ksandr. pp.40. [339.]
    1959. pp.84. [677.]
in progress.

JOHN H[OOPER] HARVEY, Conservation of old buildings. A select bibliography. University college: Bartlett school of architecture: 1958. pp.9. [100.]★

—— [another edition]. Liverpool [printed] [1959]. pp.12. [100.]

Conservation

HARRY MOSKOWITZ and JACK ROBERTS, A conservasion bibliography. Department of the army: Army library: Washington 1959. pp.iii.21. [200.]*

PAUL GAUDEL, Bibliographie der archäologischen konservierungstechnik. Berliner blätter für vor- und frühgeschichte (vol.8): Berlin 1960. pp.212. [1000.]

CONSERVATION teaching aids. A bibliography of books, bulletins and visual aids. Department of conservation: Office of information and education: [Lansing] 1961. ff.[i].7. [75.]*

Entomology.

1. *Periodicals*

SERIAL publications examined. Commonwealth

institute of entomology: 1953. pp.25. [1250.]
an earlier edition appeared in vol.xxxv of the
Review of applied entomology (*1948*).
— [another edition]. 1960. pp.[ii].16. [1000.]

LIST of serial publications in the library of the
Department of entomology. British museum
(natural history): 1958 [1959]. ff.109. [750.]★
— Second edition. 1962. ff.144. [1100.]★

2. *General*

[JEAN] CHARLES [EMMANUEL] NODIER, Bibliographie entomologique, ou catalogue raisonné des
ouvrages relatifs à l'entomologie et aux insectes.
An IX. [1800]. pp.viii.64. [400.]

JOHANN NEPOMUK EISELT, Geschichte, systematik
und literatur der insectenkunde, von den ältesten
zeiten bis auf die gegenwart. Leipzig 1836. pp.
viii.250. [1500.]

A[CHILLE] PERCHERON, Bibliographie entomologique, comprenant l'indication . . . 1º des
ouvrages entomologiques publiés en France et à
l'étranger depuis les temps les plus reculés jusques
et y compris l'année 1834; 2º des monographies et
mémoires contenus dans les recueils, journaux et
collections académiques françaises et étrangères.
1837. pp.xii.326+[iii].376. [5000.]

a copy containing ms. additions is in the British museum.

BERICHT über die wissenschaftlichen leistungen im gebiete der entomologie während des jahres 1838[–1913].
offprinted with minor changes of title, from the Archiv für naturgeschichte, *which is entered under* Zoology, *below.*

ÅRSBERÄTTELSE [*afterwards:* Berättelse] om framstegen i insekternas, myriapodernas och arachnidernas naturalhistoria. Kongliga vetenskapsakademien: Stockholm.

>1840–1842. Af C[arl] H. Boheman. pp.[iii]. iv. 150. [600.]
>1853–1854. pp.viii.295. [1000.]
>1855–1856. pp.viii.362. [1500.]

JOHANNES [VON NEPOMUK FRANZ XAVER] GISTEL, Lexikon der entomologischen welt, der carcinologischen und arachnologischen. Stuttgart 1846. pp.328. [10,000.]

BIBLIOTHECA stephensiana; being a catalogue of the entomological library of the late James Francis Stephens. 1853. pp.48. [768.]

HERMANN AUGUST HAGEN, Bibliotheca entomologica. Die litteratur über das ganze gebiet der

entomologie bis zum jahre 1862. Leipzig 1862–
1863. pp.xii.566+[ii].542. [22,500.]

—— [supplement]. Albert Müller, Contributions to entomological bibliography up to 1862.

ii. 1873. pp.15. [100.]

iii. 1873. pp.16. [100.]

the first number appears in the Transactions *of the*
Entomological society (1873), ii.207–217.

CATALOGUE of works in the library of the
American entomological society. Philadelphia
1868. pp.[ii].32. [600.]

[ALFRED PREUDHOMME DE BORRE], Catalogue de
la bibliothèque de la Société entomologique de
Belgique. [Brussels 1871–1881]. pp.[548].
[10,000.]

THE ZOOLOGICAL RECORD [XIII; *afterwards:* XI;
XII]. Insecta.

xiii. 1876. By E[dward] C[aldwell] Rye [*and*
others]. pp.240. [500.]

xiv. 1877. pp.234. [500.]

xv. 1878. pp.281. [600.]

xvi. 1879. By W[illiam] F[orsell] Kirby [and
Robert McLachlan]. pp.250. [500.]

xvii. 1880. pp.238. [500.]

xviii. 1881. pp.303. [600.]

xix. 1882. pp.292. [600.]

xx. 1883. pp.299. [600.]
xxi. 1884. pp.319. [500.]
xxii. 1885. By D[avid] Sharp [and R. McLachlan]. pp.264. [375.]
xxiii. 1886. pp.330. [751.]
xxiv. 1887. By D. Sharp. pp.323. [810.]
xxv. 1888. pp.327. [941.]
xxvi. 1889. pp.320. [949.]
xxvii. 1890. pp.320. [927.]
xxviii. 1891. pp.311. [974.]
xxix. 1892. pp.332. [1026.]
xxx. 1893. pp.371. [1069.]
xxxi. 1894. pp.384. [1173.]
xxxii. 1895. pp.387. [1251.]
xxxiii. 1896. pp.324. [1264.]
xxxiv. 1897. pp.300. [1205.]
xxxv. 1898. pp.295. [1344.]
xxxvi. 1899. pp.276. [1241.]
xxxvii. 1900. pp.354. [1431.]
xxxviii. 1901. pp.309. [1525.]
xxxix. 1902. pp.313. [1512.]
xl. 1903. pp.373. [1710.]
xli. 1904. pp.362. [1549.]
xlii. 1905. pp.342. [1669.]
xliii. 1906. pp.455. [3000.]
xliv. 1907. pp.423. [3153.]
xlv. 1908. pp.435. [3194.]

xlvi. 1909. pp.428. [3238.]
xlvii. 1910. pp.459. [3232.]
xlviii. 1911. pp.413. [2800.]
xlix. 1912. pp.459. [3255.]
l. 1913. pp.470. [2967.]
li. 1914. pp.342. [2230.]
lii. 1915. pp.276. [1972.]
liii. 1916. pp.264. [1821.]
liv. 1917. pp.227. [1447.]
lv. 1918. pp.246. [1677.]
lvi. 1919. pp.271. [1893.]
lvii. 1920. pp.289. [1932.]
lviii. 1921. By N. D. Riley. pp.270. [1970.]
lix. 1922. By the Imperial bureau [*afterwards:*
    institute] of entomology. pp.432. [3105.]
lx. 1923. pp.347. [2593.]
lxi. 1924. pp.317. [2423.]
lxii. 1925. pp.421. [3230.]
lxiii. 1926. pp.464. [3096.]
lxiv. 1927. pp.418. [2923.]
lxv. 1928. pp.430. [2882.]
lxvi. 1929. pp.436. [2955.]
lxvii. 1930. pp.435. [3024.]
lxviii. 1931. pp.436. [3132.]
l.ix. 1932. pp.446. [3238.]
lxx. 1933. pp.443. [3235.]
lxxi. 1934. pp.488. [3405.]

lxxii. 1935. pp.440. [3424.]
lxxiii. 1936. pp.445. [3725.]
lxxiv. 1937. pp.400. [3515.]
lxxv. 1938. pp.427. [3459.]
lxxvi. 1939. pp.420. [3277.]
lxxvii. 1940. pp.263. [1827.]
lxxviii. 1941. pp.242. [1769.]
lxxix. 1942. pp.206. [1594.]
lxxx. 1943. pp.235. [1526.]
lxxxi. 1944. pp.247. [1727.]
lxxxii. 1945. pp.351. [2457.]
lxxxiii. 1946. pp.380. [2857.]
lxxxiv. 1947. pp.400. [3082.]
lxxxv. 1948. pp.350. [2807.]
lxxxvi. 1949. pp.398. [3215.]
lxxxvii. 1950. pp.447. [3101.]
lxxxviii. 1951. pp.464. [3335.]
lxxxix. 1952.
xc. 1953. pp.534. [4024.]
xci. 1954. pp.546. [4060.]
xcii. 1955. pp.535. [4005.]
xciii. 1956. pp.536. [3444.]
xciv. 1957. pp.576. [3877.]
xcv. 1958. pp.660. [4521.]
xcvi. 1959. pp.384.
xcvii. 1960. pp.357.
xcviii. 1961. pp.396. [3894.]

Entomology

in progress; the earlier issues were not published separately; the issues for 1906–1914 also form part of vols.vi–xiv of the International catalogue of scientific literature, N. Zoology.

BERICHT über die wissenschaftlichen leistungen im gebiete der entomologie. Berlin.

  1880. Von Philipp Bertkau. pp.[iv].256. [3000.]

  1881. pp.[iii].298. [4000.]

  1882. pp.[iii].292. [4000.]

  1883. pp.iv.266. [3500.]

  1884. Von P. Bertkau und E[duard] von Martens, pp.iv.272. [3500.]

  1885. Von P. Bertkau. pp.i.328. [4500.]

  1886. Von P. Bertkau und G. H. Fowler. pp.[iii].387. [5000.]

  1887. Von P. Bertkau. pp.[iii].228. [5000.]

  1888. Von P. Bertkau und G. H. Fowler. pp.[iii].290. [6000.]

  1889. Von P. Bertkau. pp.[ii].318. [6000.]

  1890. Von P. Bertkau und F[ranz] Hilgendorf. pp.[iii].419. [7500.]

  1891. pp.[iii].398. [7500.]

  1892. pp.[iv].416. [7500.]

  1893. pp.[iv].416. [7500.]

no more published.

G[EORGE] C. CHAMPION, Catalogue of the library of the Entomological society of London. 1893. pp.[iv].312. [6000.]

—— Supplementary catalogue. 1900. pp.[iv]. 147. [3000.]

ENTOMOLOGISCHE litteraturblätter. Repertorium der neuesten arbeiten auf dem gesammtgebiet der entomologie. Berlin.

  i. 1901. Herausgegeben von R. Friedländer & Sohn. pp.[ii].210. [2000.]
  ii. 1902. pp.[iv].232. [2500.]
  iii. 1903. pp.[ii].254. [2500.]
  iv. 1904. pp.[iv].232. [2500.]
  v. 1905. pp.[iv].204. [2000.]
  vi. 1906. pp.[iv].208. [2000.]
  vii. 1907. pp.[iv].216. [2000.]
  viii. 1908. pp.iv.252. [2500.]
  ix. 1909. pp.iv.244. [2500.]
  x. 1910. pp.iv.252. [2500.]
  xi. 1911. pp.iv.244. [2500.]
  xii. 1912. pp.iv.268. [3000.]
  xiii. 1913. pp.iv.268. [3000.]
  xiv. 1914. pp.iv.256. [2500.]
  [continued as:]

Repertorium entomologicum. . . . Deutsche entomologische gesellschaft.

i. 1924. Bearbeitet von H. Hedicke. pp.[ii].
120. [2000.]

ii. 1925. pp.[ii].144. [2500.]

iii. 1926–1927. pp.iv.180. [3000.]

iv. 1927–1928. pp.iv.180. [3000.]

v. 1928–1929. pp.iv.176. [3000.]

vi. 1929–1930. pp.iv.184. [3000.]

vii. 1930–1931. Bearbeitet von H. Hedicke
und N. Mallach. pp.iv.184. [3000.]

viii. 1931–1932. pp.iv.180. [3000.]

ix. 1932–1934. pp.iv.176. [3000.]

x. 1934–1935. pp.iv.144. [2000.]

no more published.

каталогъ библіотеки Русскаго энтомоло-
гическаго общества. Выпускъ 1-й. Ино-
странный отдѣлъ. Изданія не періодическія. С.-Петербургъ 1903. pp.[ii].193. [2750.]
no more published?

[JOSEPHINE A. CLARK], Catalogue of publications relating to entomology in the library of the
U.S. Department of agriculture. Department of
agriculture: Library bulletin (no.55): Washington 1906. pp.562. [5600.]

KATALOG der Bibliothek des Deutschen entomologischen museums. Teil 1: Einzelwerke und

separata. Berlin 1913. pp.[ii].313. [16,500.]
no more published.

WALTER HORN and SIGM[UND] SCHENKLING,
Index litteraturae entomologicae. Serie 1: Die
weltliteratur über die gesamte entomologie bis
inklusive 1863. Berlin 1928–1929. pp.[ii].xxi.352
+ [ii].353–704 + [ii].705–1056 + [ii].1057–1426.
[25,229.]
no more published.

FINNISH entomological literature published in
1934 [&c.] including economic entomology and
control of insect pests. [Helsinki].
1934. [By] Niilo A. Vappula. pp.17. [200.]
1935. pp.22. [350.]
1936. pp.24. [350.]
1937. pp.24. [350.]
in progress.

W[ILLAR] J[OSEPH] CHAMBERLIN, Entomological
nomenclature and literature. Ann Arbor 1941.
pp.ix.103. [500.]★
—— 3rd edition. Dubuque [1952]. pp.vii.141.
[1250.]★

3. *Agricultural and applied entomology*

i. *Periodicals*

J[OHANNES] BÄRNER, Literaturquellen und ihre

kürzungen aus der bibliographie der pflanzenschutzliteratur. Biologische bundesanstalt für
land- und forstwirtschaft: Berlin 1958. ff.[ii].167.
[1750.]*

ii. *General*

[v. AUDOIN], Notice sur les recherches d'entomologie agricole de m. V. Audoin. 1838. pp.22.
[63.]

SAMUEL HENSHAW [parts vi–viii: NATHAN
BANKS], Bibliography of the more important contributions to American economic entomology.
U.S. Department of agriculture: Division of entomology: Washington [1889–]1890–1905. pp.[ii].
454+167+179+273+113+132. [12,500.]

— — Index to the literature of american economic entomology, January 1, 1905 to December 31,
1914. Compiled by Nathan Banks. American
association of economic entomologists: Special
publication (no.1): Melrose Highlands, Mass.
1917. pp.[ii].v.323. [20,000.]

— — Index II . . . 1915 to . . . 1919. Compiled
by Mabel Colcord. Edited by E. Porter Felt . . .
(no.2): 1921. pp.[viii].388. [22,500.]

— — Index III . . . 1920 to . . . 1924 . . . (no.3):
1925. pp.[x].441. [25,000.]

— — Index IV . . . 1925 to . . . 1929 . . . (no.4):
1930. pp.[xi].518. [30,000.]

—— Index v . . . 1930 to . . . 1934 . . . (no.5):
1938. pp.[xii].693. [42,500.]

G. W. KIRKALDY, A bibliography of sugar-cane
entomology. Hawaiian sugar planters' association:
Experiment station: Division of entomology:
Bulletin (no.8): Honolulu 1909. pp.74. [350.]

THE REVIEW of applied entomology. Series A:
Agricultural. Imperial [*afterwards:* Common-
wealth] bureau [*afterwards:* institute] of entomology.

   i. 1913. pp.[vii].629. [750.]
   ii. 1914. pp.viii.794. [1000.]
   iii. 1915. pp.viii.865. [1000.]
   — Subject index to vols.I–III. Compiled by
     S. A. Neave. 1917. pp.[iii].332.
   iv. 1916. pp.viii.704. [1000.]
   v. 1917. pp.iv.767. [1000.]
   vi. 1918. pp.iii.750. [1000.]
   vii. 1919. pp.iv.728. [1000.]
   viii. 1920. pp.iv.742. [1000.]
   ix. 1921. pp.iv.831. [1250.]
   x. 1922. pp.iv.845. [1250.]
   xi. 1923. pp.iv.790. [1250.]
   xii. 1924. pp.iii.780. [1250.]
   xiii. 1925. pp.iv.845. [1500.]
   xiv. 1926. pp.iv.839. [1500.]

xv. 1927. pp.[iii].854. [1500.]
xvi. 1928. pp.[iii].864. [1500.]
xvii. 1929. pp.[iii].910. [1750.]
xviii. 1930. pp.[iii].895. [1750.]
xix. 1931. pp.[iii].951. [2000.]
xx. 1932. pp.[iii].903. [2000.]
xxi. 1933. pp.[iii].862. [1750.]
xxii. 1934. pp.[iii].918. [2000.]
xxiii. 1935. pp.[iv].950. [2000.]
xxiv. 1936. pp.[iii].1011. [2000.]
xxv. 1937. pp.[iii].996. [2000.]
xxvi. 1938. pp.[iii].956. [2000.]
xxvii. 1939. pp.[iii].860. [1750.]
xxviii. 1940. pp.[iii].814. [1500.]
xxix. 1941. pp.[iii].816. [1500.]
xxx. 1942. pp.[iii].750. [1500.]
xxxi. 1943. pp.[iii].652. [1300.]
xxxii. 1944. pp.[iii].538. [1100.]
xxxiii. 1945. pp.[ii].503. [1000.]
xxxiv. 1946. pp.[iii].483. [1000.]
xxxv. 1947. pp.[iii].562. [1100.]
xxxvi. 1948. pp.[iii].527. [1000.]
xxxvii. 1949. pp.[iii].591. [1200.]
xxxviii. 1950. pp.[iii].611. [1200.]
xxxix. 1951. pp.[iii].557. [1100.]
xl. 1952. pp.[iii].492. [1000.]
xli. 1953. pp.[iii].548. [1000.]

xlii. 1954. pp.[iii].513. [1000.]
xliii. 1955. pp.[iii].552. [1000.]
xliv. 1956. pp.[iii].569. [1000.]
xlv. 1957. pp.[iii].615. [1250.]
xlvi. 1958. pp.[iii].624. [1250.]
xlvii. 1959. pp.[iii].627. [1250.]
xlviii. 1960. pp.[iii].   . [1500.]
xlix. 1961. pp.[iv].689. [1750.]
l. 1962. pp.[iii].842. [2000.]
li. 1963. pp.[iii].872. [2000.]
in progress; series B is entered in section 3 below.

BIBLIOGRAPHIE der pflanzenschutzliteratur. Biologische reichsanstalt für land- und forstwirtschaft: Berlin.

1914–1919. Bearbeitet von H[ermann] Morstatt. pp.viii.464. [10,000.]
1920. ff.71. [1500.]
printed on one side of the leaf.

1921. pp.198. [4500.]
1922. pp.iv.162. [3500.]
1923. pp.iv.176. [4000.]
1924. pp.iv.226. [5000.]
1925. pp.iv.228. [5000.]
1926. pp.iv.231. [5000.]
1927. pp.iv.250. [5500.]
1928. pp.iv.251. [5500.]

1929. pp.iv.246. [5500.]
1930. pp.iv.245. [5500.]
1931. pp.iv.251. [5500.]
1932. pp.iv.259. [5500.]
1933. pp.iv.316. [7500.]
1934. pp.iv.302. [7000.]
1935. pp.iv.352. [8000.]
1936. pp.iv.392. [9000.]
1937. pp.iv.430. [9000.]
1938.
1939. pp.iv.378. [7000.]
no more published.

N. N. BOGDANOV-KATKOV, Список русской литературы по прикладной энтомологии за 1917–1921 г. г. Сельско-хозяйственный ученый комитет: Отдел прикладной энтомологии. Петроград 1921. pp.[ii].16. [300.]

N. N. BOGDANOV-KATKOV, Русская литература по прикладной энтомологии (преимущественно сельско-хозяйственная). Ленинград 1924. pp.xii.224. [5000.]

MABEL COLCORD, INA L[OUISE] HAWES and ANGELINA J[ACQUELINE] CARABELLI, Check list of publications on entomology issued by the United States Department of agriculture through 1927.

Entomology

Department of agriculture: Bibliographical contributions (no.20): Washington 1930. pp.iv.261. [3000.]★

J[OSEPH] S[ANFORD] WADE, List of entomological publications of personnel of cereal and forage insect investigations, U.S. Bureau of entomology, 1904–1928, inclusive. [Washington 1930]. ff.46. [700.]

PUBLICATIONS of the Insecticide division. Department of agriculture: Bureau of chemistry and soils: [Washington 1931]. ff.9. [112.]★
— [another edition]. List of publications of the Division of insecticide investigations [1947]. pp.80. [923.]★
irregular supplements have been issued.

LIST of publications of the Pesticide chemicals research section [branch]. Department of agriculture: Agricultural research administration: Entomological research branch [division]: [Beltsville] [?1934] &c.★
in progress.

WALKER E[UGENE] MCBATH, A bibliography on the use of airplanes in insect control from 1922 to 1933. Bureau of entomology: [Washington 1934]. pp.[ii].37. [175.]★

Entomology

RECENT literature on the toxicity and insecticidal value of rotenone. Science library: Bibliographical series (no.135): 1934. ff.[i].4. [71.]★
— Supplement . . . (no.240): 1936. ff.8. [146.]★

THE USE of aeroplanes in the control of pests. Science library: Bibliographical series (no.278): 1936. ff.16. [278.]★

MISCELLANEOUS references on insecticides and fish-poisons of vegetable origin. Science library: Bibliographical series (no.317): 1937. ff.14. [250.]★

F. E. DEARBORN, The arsenates of manganese as insecticides. (A review of the literature). Bureau of entomology and plant quarantine: Division of insecticide investigations: Washington 1937. pp. 2–27. [89.]★

R. C. ROARK, Tephrosia as an insecticide — a review of the literature. Bureau of entomology and plant quarantine: Division of insecticide investigations: Washington 1937. pp.165. [601.]★

F. E. DEARBORN, The arsenates of magnesium as insecticides. (A review of the literature). Bureau of entomology and plant quarantine: Division of insecticide investigations: Washington 1938. pp.42. [161.]★

C. E. PETCH, Index to entomological publications of the Department of agriculture, 1884–1936. Ottawa 1938 &c.★

in progress.

CARL J[OHN] DRAKE, Bibliography of the state plant pest laws, quarantines, regulations and administrative rulings of the United States of America. National plant board: [*s.l.* 1943]. ff.[ii].60. [300.]

BIBLIOGRAPHY on insect pest resistance in plants (with a supplement on resistance to nematodes). Imperial bureau of plant breeding and genetics: School of agriculture: Cambridge 1944. pp.40. [518.]

H[AROLD] C. GOUGH, A review of the literature on soil insecticides. Imperial institute of entomology: 1945. pp.[iii].161. [650.]

LIST of United States patents granted members of the Division of insecticide investigations for the period July 1, 1927, to June 30, 1947. Department of agriculture: Bureau of entomology and plant quarantine: [Washington 1947]. pp.24. [144.]★

J[OSEPH] S[ANFORD] WADE, A selected bibliography of the insects of the world associated with sugar cane, their predators and parasites. International society of sugar cane technologists: Memoir (no.1): Honolulu 1951. pp.113. [1000.]

FRANCIS D. HORIGAN and FLORENCE H[ARDEN] COZZI, The persistency of insecticidual [*sic*] residues on various surfaces. QM research & development laboratories: Technical library: Bibliographic series (no.23): Philadelphia [1952]. pp.[ii].23. [91.]*

ANNOTATED bibliography of analytical methods for pesticides. National research council: Publication (no.241 &c.): Washington 1952 &c.*
in progress.

[MONIKA MÜLLER], Schädlingsbekämpfung. Eine zusammenstellung der seit 1930 in deutscher sprache erschienenen selbständigen schriften. Schriften zum bibliotheks- und büchereiwesen in Sachsen-Anhalt (no.9): Leipzig 1953. pp.71. [821.]

V[ERA] V[LADIMIROVNA] GNUCHEVA, Защита сельскохозяйственных культур от вредителей, болезней и сорняков. Рекомендательный указатель литературы. Государственная . . . публичная библиотека имени М. Е. Салтыкова-Щедрина: Ленинград 1955. pp.28. [50.]

J[OHANNES] BÄRNER [*and others*], Internationale

pflanzenschutzliteratur. Berlin [1957]. pp.48. [383.]

4. *Medical and veterinary entomology*

J[OHANN] CH[RISTOPH] HUBER, Bibliographie der klinischen entomologie (hexapoden, acarinen). Jena 1899–1900. pp.[ii].24+[ii].24+[ii].25+ [ii].27. [1000.]

THE REVIEW of applied entomology. Series B: Medical and veterinary. Imperial [*afterwards:* Commonwealth] bureau [*afterwards:* institute] of entomology.

   i. 1913. pp.[v].276. [500.]
   ii. 1914. pp.[vii].249. [500.]
   iii. 1915. pp.[vii].288. [500.]
   iv. 1916. pp.[vii].248. [500.]
   v. 1917. pp.[iii].237. [500.]
   vi. 1918. pp.[iii].291. [500.]
   vii. 1919. pp.[iii].239. [500.]
   viii. 1920. pp.[iii].280. [500.]
   ix. 1921. pp.[iii].284. [500.]
   x. 1922. pp.[iii].307. [750.]
   — Index to vols.i–x. 1924. pp.[iii].219.
   xi. 1923. pp.[iii].286. [750.]
   xii. 1924. pp.[iii].256. [750.]
   xiii. 1925. pp.[iii].248. [750.]
   xiv. 1926. pp.[iii].282. [750.]

xv. 1927. pp.[iii].292. [750.]
xvi. 1928. pp.[iii].320. [1000.]
xvii. 1929. pp.[iii].307. [1000.]
xviii. 1930. pp.[iii].331. [1000.]
xix. 1931. pp.[iii].315. [1000.]
xx. 1932. pp.[iii].338. [1000.]
xxi. 1933. pp.[iii].342. [1000.]
xxii. 1934. pp.[iii].318. [1000.]
xxiii. 1935. pp.[iii].369. [1000.]
xxiv. 1936. pp.[iii].367. [1000.]
xxv. 1937. pp.[iii].343. [1000.]
xxvi. 1938. pp.[iii].310. [1000.]
xxvii. 1939. pp.[iii].320. [1000.]
xxviii. 1940. pp.[iii].306. [900.]
xxix. 1941. pp.[iii].246. [750.]
xxx. 1942. pp.[ii].238. [750.]
xxxi. 1943. pp.[ii].300. [900.]
xxxii. 1944. pp.[ii].284. [900.]
xxxiii. 1945. pp.[ii].248. [750.]
xxxiv. 1946. pp.[ii].252. [750.]
xxxv. 1947. pp.[ii].286. [750.]
xxxvi. 1948. pp.[iii].262. [750.]
xxxvii. 1949. pp.[iii].290. [750.]
xxxviii. 1950. pp.[iii].270. [750.]
xxxix. 1951. pp.[iii].263. [750.]
xl. 1952. pp.[iii].266. [750.]
xli. 1953. pp.[iii].261 [750.]

xlii. 1954. pp.[iii].244. [750.]
xliii. 1955. pp.[iii].254. [750.]
xliv. 1956. pp.[iii].248. [750.]
xlv. 1957. pp.[iii].253. [750.]
xlvi. 1958. pp.[iii].259. [750.]
xlvii. 1959. pp.[ii].251. [750.]
xlviii. 1960. pp.[iii].291. [750.]
xlix. 1961. pp.[iii].363. [1000.]
l. 1962. pp.[iii].384. [1000.]
li. 1963. pp.[iii].360. [1000.]

in progress; series A is entered in section 2 above.

KENNETH M. HUGHES, An annotated list and bibliography of insects reported to have virus diseases. Hilgardia (vol.xxvi, no.14): Berkeley 1957. pp.597–629. [255.]

5. Countries

America, north

[CHARLES V. RILEY], An enumeration of the published synopses, catalogues and lists of north american insects. Department of agriculture: Division of entomology: Bulletin (no.19): Washington 1888. pp.77. [800.]

NATHAN BANKS, A list of works on north american entomology. Department of agriculture: Division of entomology: Washington 1900. pp.

95. [1250.]
— [another edition] 1910. pp.120. [1750.]

Argentina

CARLOS LIZER, Primer ensayo bibliográfico de entomología argentina. Buenos Aires 1919. pp. [ii].351–380. [356.]

AUGUSTO A. PIRAN, Bibliografía entomológica argentina, 1847–1952. Sociedad entomológica argentina: Curso de entomología (vol.x): Buenos Aires 1954. pp.[v].623–716. [2000.]
A–H only; no more published.

Australia

ANTHONY MUSGRAVE, Bibliography of australian entomology, 1775–1930. Royal zoological society of New South Wales: Sydney 1932. pp. viii.380. [6000.]

China

YOSHO ÔUCHI, Bibliographical introduction to the study of chinese insects. Shanghai science institute: Department of biology: Studies (Entomological report, no.1): Shanghai 1934. pp.[v]. 533. [3500.]

also forms vol.ii of section iii of The journal of the Shanghai science institute.

Entomology

Great Britain

JAMES FRANCIS STEPHENS, A systematic catalogue of british insects. . . . Containing also the references to every english writer on entomology, and to the principal foreign authors. 1829. pp.xxxv. 416.388. [25,000.]

GEORGE SIDNEY KLOET and WALTER DOUGLAS HINCKS, A check list of british insects. Stockport 1945. pp.lix.483.[xiii]. [35,000.]

Hawaii

J[AMES] F[RANKLIN] ILLINGWORTH, Early references to hawaiian entomology. Bernice P. Bishop museum: Bulletin (no.2): Honolulu 1923. pp.63. [225.]

India

LIST of publications on indian entomology. Agricultural research institute, Pusa: Bulletin (no.139 &c.): Bombay [1930 &c.: Imperial council of agricultural research: Miscellaneous bulletin (no.1, &c.): Delhi].

 1920–1921. [By T. B. Fletcher] . . . (no.139): 1922. pp.[ii].67. [750.]
 1922 . . . (no.147): 1923. pp.[ii].42. [500.]
 1923 . . . (no.155): 1924. pp.[ii].59. [750.]
 1924 . . . (no.161): 1925. pp.[ii].41. [500.]

Entomology

1925 ... (no.165): 1926. pp.[ii].62.x. [750.]
1926 ... (no.168): 1927. pp.[ii].48. [500.]
1927 ... (no.184): 1928. pp.[ii].33. [400.]
1928 ... (no.200): 1929. pp.[ii].33. [400.]
1929 ... (no.207): 1931. pp.[ii].36. [400.]
1930 ... (no.1): 1934. pp.[ii].45. [500.]
1931 ... (no.2): 1934. pp.[ii].27. [300.]
1932. [By T. B. Fletcher and P. V. Isaac]. ...
   (no.3): 1934. pp.[ii].36. [400.]
1933. [By P. V. Isaac] ... (no.5): 1935. pp.[ii].
   29. [300.]
1934 ... (no.7): 1935. pp.[ii].38. [400.]
1935. [By H. S. Pruthi] ... (no.14): 1937.
   pp.[ii].40. [400.]

Micronesia

E[DWIN] H[ORACE] BRYAN, Bibliography of micronesian entomology. National research council: Pacific science board: [s.l.] 1943. ff.43. [500.]*

New Zealand

DAVID MILLER, Bibliography of New Zealand entomology, 1875–1952 (with annotations). Department of scientific and industrial research: Bulletin (no.120): Wellington 1956. pp.iii–xxiv. 493. [3500.]

the year 1875 on the titlepage should read 1775.

Entomology

Puerto Rico

MORTIMER D. LEONARD, An annotated bibliography of puerto rican entomology. Insular experiment station: Journal of the Department of agriculture of Puerto Rico (vol.xvii, no.1): Río Pedras 1933. pp.96. [750.]

Russia

N[IKOLAI] N[IKOLAEVICH] BOGDANOV-KATKOV, Русская литература по прикладной энтомологии (преимущественно сельско-хозяйственная). Ленинград 1924. pp.xii.224. [5000.]

[AKSEL NIKOLAEVICH REIKHARDT and ALEKSANDR ALEKSANDROVICH SHTAKELBERG], Список литературы по вредным насекомым СССР за 1925–1928 г.г. Всесоюзная академия с.-х. наук им. В. И. Ленина: Институт защиты растений (no.20): Ленинград 1931. pp.96. [2517.]

Switzerland

THEOD. STECK, Insecta, 1634–1920. Centralkommission für schweizerische landeskunde: Bibliographie der schweizerischen landeskunde: Fauna helvetica (IV. 6–7α): Bern 1926. pp.vi.292. [3000.]

Entomology

United States

ENTOMOLOGY bureau bulletins, circulars and technical series relating to insects. Superintendent of documents: Price list (no.41): Washington [1909]. pp.24. [350.]
frequently re-edited under various titles.

LIST of books on insects of the United States. Library of Congress: Washington 1918. single leaf. [8.]★

WILTON EVERETT BRITTON, Check-list of the insects of Connecticut. State geological and natural history survey: Bulletin (no.31): Hartford 1920. pp.397. [6000.]
— — Additions . . . (no.60): 1938. pp.169. [3000.]

MELVILLE H. HATCH, A bibliographical catalogue of the injurious arachnids and insects of Washington. University of Washington: Publications in biology (vol.i, no.4): Seattle 1938. pp.163–223. [1000.]

6. *Miscellaneous*

[ALEXANDRE LABOULBÈNE], Travaux d'entomologie du d^r A. Laboulbène. [1867]. pp.8. [67.]

SAMUEL HENSHAW, The entomological writings

of John L. Leconte. Dimmock's special bibliography (no.1): Cambridge, Mass. 1878. pp.[ii].11. [152.]

SAMUEL HENSHAW, The entomological writings of George H. Horn. Dimmock's special bibliography (no.2): Cambridge, Mass. 1879. pp.[ii].6. [80.]

SAMUEL H[UBBARD] SCUDDER, A bibliography of fossil insects. Library of Harvard university: Bibliographical contributions (no.13): Cambridge, Mass. 1882. pp.47. [1100.]

SAMUEL HENSHAW, The entomological writings of dr. Alpheus Spring Packard. Department of agriculture: Division of entomology: Bulletin (no.16): Washington 1887. pp.49. [339.]

SAMUEL HUBBARD SCUDDER, A classed and annotated bibliography of fossil insects. Geological survey: Bulletin (no.69): Washington 1890. pp. 101. [1000.]

A[NTON] HANDLIRSCH, Insecta palaeozoica. Fossilium catalogus (1. 16): Berlin 1922. pp.[ii].230. [6000.]

CHARLES P. ALEXANDER, An entomological survey of the Salt Fork of the Vermilion river in

1921, with a bibliography of aquatic insects. State of Illinois: Natural history survey: Bulletin (vol. xv, no.viii): Urbana 1925. pp.[iii].439–535. [1000.]

INA L[OUISE] HAWES, Bibliography of the effect of light on insects. [Bureau of entomology and plant quarantine: Washington] 1935. pp.10. [100.]*

FRANCIS HEMMING, Hübner. A bibliographical and systematic account of the entomological works of Jacob Hübner and of the supplements thereto by Carl Geyer, Gottfried Franz von Frölich and Gottlieb August Wilhelm Herrich-Schäffer. Royal entomological society: 1937. pp. xxxiv.605+xi.274. [100.]

THE RESPIRATION of insects. (References from 1930–date 11.VI.37). Science library: Bibliographical series (no.330): 1937. ff.7. [110.]*
— 1930–date 24.II.39. . . . (no.462): 1939. ff.11. [200.]*

DAS INSEKT in der darstellung vom mittelalter bis auf Linné. Ausstellung der Bayerischen staatsbibliothek. München 1938. pp.47. [258.]
limited to manuscripts and printed books.

ELECTROCUTING light traps and suction light

traps for catching insects. Science library: Bibliographical series (no.487): 1939. ff.2. [27.]*

INA L[OUISE] HAWES and ROSE EISENBERG, Bibliography on aviation and economic entomology. Department of agriculture: Bibliographical bulletin (no.8): Washington 1947. pp.iv.186. [1084.]

ACTION of high frequency fields on insects and bacteria. Science library: Bibliographical series (no.676): 1949. ff.2. [35.]*

ELLIS A[RDEN] HICKS, Check-list and bibliography of the occurrence of insects in birds' nests. Ames [1959]. pp.683. [17,500.]

MABLE and HUBERT FRINGS, Sound productions and sound reception by insects. A bibliography. [State College 1960]. pp.[vi].108. [1752.]*

INSECT respiration (excluding morphology of the respiratory system). References 1953–1959. Science library: Bibliographical series (no.775): [1960]. pp.11. [200.]

Fish, fisheries, fishing.

1. *General*

PETRUS ARTEDI, Bibliotheca ichthyologica seu historica litteraria ichthyologiæ [Edited by Carolus Linnaeus]. Lugduni Batavorum 1738. pp.[iv].68. [500.]

Fish

— — Pietri Artedi Renovati pars I. et II.[–v.] i.e. bibliotheca et philosophia ichthyologica. Cura Iohannis Iulii Walbaumii. Grypeswaldiæ [1788–] 1789–1793. pp.[viii].230 + [vi].196.[iii] + [viii]. 723+[iv].140+[iv].112. [1500.]

the first part was first issued with a temporary title-page with somewhat different wording, dated 1788.

[SIR HENRY ELLIS], A catalogue of books on angling. 1811. pp.21. [80.]

a copy in the British museum contains numerous ms. notes and additions by William White (of Crick-howell) and Joseph Haslewood.

— — [another edition]. Bibliotheca piscatoria. A catalogue of books upon angling. 1836. pp.[40]. [250.]

also issued as a supplement to [Thomas Boosey], Piscatorial reminiscences and gleanings, 1835.

[JEAN] L[OUIS RODOLPHE] AGASSIZ, Nomina systematica generum piscium. [Soloduri 1842–1846]. pp.viii.69.5. [2600.]

T[HOMAS] WESTWOOD, A new bibliotheca piscatoria; or general catalogue of angling and fishing literature. 1861. pp.78. [600.]

JOSEPH SABIN, A bibliographical catalogue of the waltonian library . . . of Robert W. Coleman.

New York 1866. pp.[ii].149. [700.]

ABRIDGMENTS of specifications. Class 48. Fish and fishing. Patent office.

    1855–1866. pp.v.18. [75.]
    1867–1876. pp.v.18. [75.]
    1877–1883. pp.iv.20. [75.]
    1884–1888. pp.vi.32. [125.]
    1889–1892. pp.vi.24. [100.]
    1893–1894. pp.vi.32. [125.]
    1895–1900. pp.vi.32. [125.]
    1901–1904. pp.vi.38. [150.]
    1905–1908. pp.viii.46. [200.]
    1909–1915. pp.vii.54. [200.]
    1916–1920. pp.vi.22. [100.]
    1921–1925. pp.vii.47. [200.]
    1926–1930. pp.vii.42. [200.]
no more published.

D[IRK] MULDER BOSGOED, Proeve van eene ichthyologische bibliographie. Nederlandsche maatschappij ter bevordering van nijverheid: Haarlem 1871. pp.[ii].v.247. [3351.]

D[IRK] MULDER BOSGOED, Bibliotheca ichthyologica et piscatoria. Haarlem 1874. pp.[iii].xxvi. 474. [6436.]

THE ZOOLOGICAL RECORD. [v; *afterwards:* xv; XIII; XIV; xv]. Pisces.

xiii. 1876. By A. W. E. O'Shaughnessy. pp. 38. [750.]

xiv. 1877. pp.30. [600.]

xv. 1878. pp.37. [750.]

xvi. 1879. pp.23. [500.]

xvii. 1880. By G[eorge] A[lbert] Boulenger. pp.23. [500.]

xviii. 1881. pp.25. [500.]

xix. 1882. pp.31. [600.]

xx. 1883. By G. A. Boulenger and W[illiam] R[obert] Ogilvie-Grant. pp.41. [600.]

xxi. 1884. By W. R. Ogilvie-Grant. pp.38. [500.]

xxii. 1885. pp.34. [400.]

xxiii. 1886. pp.24. [300.]

xxiv. 1887. By G. A. Boulenger. pp.34. [400.]

xxv. 1888. pp.30. [300.]

xxvi. 1889. pp.30. [300.]

xxvii. 1890. pp.35. [300.]

xxviii. 1891. pp.41. [300.]

xxix. 1892. pp.38. [300.]

xxx. 1893. pp.39. [300.]

xxxi. 1894. pp.44. [300.]

xxxii. 1895. pp.43. [300.]

xxxiii. 1896. pp.43. [250.]

xxxiv. 1897. pp.33. [250.]

xxxv. 1898. pp.45. [300.]

xxxvi. 1899. pp.43. [250.]

xxxvii. 1900. pp.34. [250.]

xxxviii. 1901. pp.39. [250.]

xxxix. 1902. pp.47. [300.]

xl. 1903. pp.43. [300.]

xli. 1904. pp.57. [350.]

xlii. 1905. By I[gerna] B. J. Sollas. pp.53.
  [504.]

xliii. 1906. pp.60. [500.]

xliv. 1907. By C. L. Boulenger. pp.56. [544.]

xlv. 1908. pp.51. [548.]

xlvi. 1909. pp.44. [424.]

xlvii. 1910. pp.39. [412.]

xlviii. 1911. By C[harles] Tate Regan. pp.47.
  [390.]

xlix. 1912. pp.52. [455.]

l. 1913. pp.44. [333.]

li. 1914. pp.28. [243.]

lii. 1915. pp.20. [182.]

liii. 1916. pp.26. [231.]

liv. 1917. pp.19. [153.]

lv. 1918. pp.19. [175.]

lvi. 1919. pp.18. [136.]

lvii. 1920. By J. R. Norman. pp.24. [243.]

lviii. 1921. pp.23. [238.]

lix. 1922. pp.30. [309.]

lx. 1923. pp.36. [397.]

lxi. 1924. pp.40. [432.]

lxii. 1925. pp.40. [432.]

lxiii. 1926. pp.45. [556.]

lxiv. 1927. By M. Burton. pp.46. [531.]

lxv. 1928. pp.49. [657.]

lxvi. 1929. pp.59. [798.]

lxvii. 1930. pp.64. [830.]

lxviii. 1931. pp.64. [785.]

lxix. 1932. pp.62. [814.]

lxx. 1933. pp.63. [793.]

lxxi. 1934. pp.66. [868.]

lxxii. 1935. pp.81. [1103.]

lxxiii. 1936. By E. Trewavas. pp.91. [1158.]

lxxiv. 1937. pp.99. [1175.]

lxxv. 1938. By C. T. Regan, J. R. Norman and E. I. White. pp.61. [714.]

lxxvi. 1939. By E. Trewavas and E. I. White. pp.85. [1098.]

lxxvii. 1940. By E. Trewavas. pp.47. [521.]

lxxviii. 1941. pp.46. [516.]

lxxix. [*not published.*]

lxxx. 1942–1943. pp.79. [837.]

lxxxi. 1944. By E. Trewavas and E. I. White. pp.62. [614.]

lxxxii. 1945. By E. Trewavas, N. B. Marshall and E. I. White. pp.59. [637.]

lxxxiii. 1946. pp.61. [630.]

lxxxiv. 1947. By N. B. Marshall, G. Palmer
and E. I. White. pp.65. [719.]

lxxxv. 1948. pp.75. [850.]

lxxxvi. 1949. pp.80. [868.]

lxxxvii. 1950. By G. Palmer and E. I. White.
pp.83. [1082.]

lxxxviii. 1951. pp.85. [1088.]

lxxxix. 1952. pp.94. [1115.]

xc. 1953. pp.90. [1133.]

xci. 1954. pp.98. [1158.]

xcii. 1955. pp.96. [1136.]

xciii. 1956. pp.93. [1100.]

xciv. 1957. pp.88. [1100.]

xcv. 1958. pp.141. [1750.]

xcvi. 1959. pp.81. [1500.]

xcvii. 1960. pp.83. [1750.]

xcviii. 1961. pp.83. [1750.]

*in progress; earlier issues were not published sepa-
rately; the issued for 1906–1914 also form part of
vols.vi–xiv of the* International catalogue of scientific literature, N. Zoology.

[JOHN BARTLETT], Catalogue of books on
angling, including ichthyology, pisciculture, fisheries, and fishing laws. From the library of a
practitioner of more than fifty years' experience

in the art of angling. Cambridge [Mass.] 1882.
pp.[ii].77. [700.]

[—] — Supplement. 1886. pp.[ii].24. [250.]
*the collection was afterwards presented to Harvard
university; see 1896, below.*

J[OHN] J[ACKSON] MANLEY, Literature of sea and
river fishing. International fisheries exhibition:
1883. pp.vii.160. [500.]

T[HOMAS] WESTWOOD and T[HOMAS] SATCHELL,
Bibliotheca piscatoria. A catalogue of books on
angling, the fisheries and fish-culture. 1883. pp.
xxiv.398. [4000.]

— — A list of books relating to fish, fishing,
and fisheries to supplement the Bibliotheca piscatoria. [By Robert Bright Marston]. English catalogue ([1900], appendix C): 1901. pp.749–776.
[1250.]

KATALOG der bibliothek des Oesterreichischen
fischerei-vereines. Wien 1892. pp.16. [200.]

[SAMUEL AUSTIN ALLIBONE], The Waltonian collection. Lenox library: Contributions to a catalogue (no.vii): New York 1893. pp.56. [456.]

R[OBERT] B[RIGHT] MARSTON, Walton and some
earlier writers on fish and fishing. Book-lover's

library: 1894. pp.xxvii.264. [50.]
reissued in 1903.

LOUISE RANKIN ALBEE, The Bartlett collection.
A list of books on angling, fishes, and fish culture,
in Harvard college library. Library of Harvard
university: Bibliographical contributions (no.51):
Cambridge, Mass. 1896. pp.180. [1300.]
see 1882–1886, above.

A CATALOGUE of the collection of books on
angling belonging to mr. Dean Sage. New-York
1896. pp.xi.245. [2000.]
50 copies printed.
— A supplement. 1904. pp.[iii].24. [200.]

[JOSEPH WILLIAM ZAEHNSDORF], Catalogue of
the books and pictures belonging to the Piscatorial
society. 1902. ff.[113]. [1000.]*
*the copy in the British museum contains supple-
mentary entries.*

H[ANS FRIEDRICH WILHELM] LICHTENFELT, Literatur zur fischkunde. Eine vorarbeit. Bonn 1906.
pp.viii.140. [2500.]

LIST of works in the library relating to fishing
and fish culture. Public library: New York 1909.
pp.49. [1500.]

Fish

A CATALOGUE of an exhibition of angling books, together with a number of manuscripts. Grolier club: New York 1911. pp.viii.60. [165.]

the collection was presented to Harvard university.

CATALOGUS van de bibliotheek. Rijksinstituten voor visscherijonderzoek: Helder 1913. pp.xi.88. [1100.]

— Lijst der aanwinsten. 1917. pp.44. [500.]

PATENTS for inventions. Fifty years subject index, 1861–1910. Class 48, fish and fishing. Patent office: 1913. pp.7. [750.]

THE ATHENÆUM subject index to periodicals, [1915]. . . . Sports and games (including fisheries). [1916]. pp.7. [250.]

no more sections devoted to fisheries were published.

BASHFORD DEAN, A bibliography of fishes. . . . Enlarged and edited by Charles Rochester Eastman. American museum of natural history: New York 1916–1923. pp.xii.718+[vi].702+xvi.707. [45,000.]

— — [another edition]. [Vol.iii edited by Eugene Willis Gudger and Arthur Wilbur Henn]. 1962. pp.xii.718+[viii].702+xvi.707. [50,000.]

350 copies printed.

CATALOGUE de la bibliothèque de pêche de m. G. Albert Petit. 1921. pp.[iii].291. [2220.]

Fish

PRICED catalogue of the collection of books on angling formed by A. E. Wilson-Browne. Birmingham [printed] 1924. pp.73. [1500.]

ANGLING. A selected list of books compiled from 'A classified list of angling books'. Revised edition. [National book council]: Bibliography (no.7): 1927. pp.[2]. [50.]

J[OHN] FITZGERALD HAMPTON, Modern angling bibliography. Books published on angling, fisheries, fish culture, from 1881 to 1945. 1947. pp.99. [1100.]

COMMERCIAL fisheries abstracts. Fish and wildlife service: Branch of commercial fisheries: Washington 1948 &c.
in progress.

W. H. SWAN, A guide to the books in the library of the C.S.A.S. [Civil service angling society]. [1949]. pp.4. [54.]*

CATALOGUS. Directie van de visserijen: Bibliotheek: [The Hague] 1951. pp.[ix].263 [5000.]*

READERS' guide to fishing and angling. Library association: County libraries section: [Readers guide (new series, no.16): 1952]. pp.27. [350.]

BIBLIOGRAFÍA de tecnología pesquera. Organización de las Naciones unidas para la agricultura y la alimentación: Extractos de la pesca mundial (vol.3, supplement): Roma 1952. pp.48. [50.]

M[AURICE] E[ARL] STANSBY, Thirty-five year index of technological publications [of the fishery technological publications of interest to fishery technologists]. Fish and wildlife service: Seattle 1953. ff.[91]+[195]+[200]+[i].51. [2000.]*

—— [another edition]. Index of fishery technological publications of the Fish and wildlife service and the former Bureau of fisheries, 1918–1955. Fish and wildlife service: Circular (no.96): 1961. pp.ii.237. [4750.]

BIBLIOGRAPHIE, l'industrie des pêches. Muséum national d'histoire naturelle: Laboratoire des pêches coloniales: 1954. ff.37. [275.]*

ARTHUR RANSOME, Fishing. National book league: Reader's guides (2nd ser., no.2): Cambridge 1955. pp.31. [100.]

CURRENT bibliography for aquatic sciences and fisheries. Food and agriculture organization: Fisheries division: Biology branch: Rome.

    iii–iv. 1960. pp.[iv].808+653.[230]+[209]. [9999.]

    v. [1961–1962]. pp.1050.[161]. [6866.]

    vi. 1963. pp.1001.[168]. [6846.]

in progress; the first two volumes were issued in a duplicated and confused form.

Fish

2. Countries

Asia

N. S. ROMANOV, Указатель литературы по рыбному хозяйству дальнего востока за 1923–1956 гг. Академия наук СССР: Отделение биологических наук: Ихтиологическая комиссия: Москва 1959. pp.292. [3690.]

Canada

LIST of references on the arbitration of Canadian fisheries disputes. Library of Congress: Washington 1920. ff.19. [172.]*

JACQUES SANSFAÇON and VIANNEY LEGENDRE, Bibliographie des titres des documents sériés ayant publié [sic] sur les poissons, la pêche et les pêcheries du Canada. Édition préliminaire. Province de Québec: Ministère de la chasse et des pêcheries: Office de biologie [&c.]: [s.l.] 1957. pp.v.xx.106. [641.]

GENERAL information on lake and stream maps. Michigan department of conservation: Fish division: Pamphlet (no.17): Lansing 1958. pp.[175]. [2000.]*

Czechoslovakia

DAGMAR HOFMANOVÁ, Za další rozvoj našeho

Fish

rybářství. Bibliografie z fondů vědeckého oddělení Krajské knihovny. České Budějovice 1959. pp.32. [216.]

Germany

VERZEICHNISS sämmtlicher schriften über fischerei, fischzucht etc. welche von 1820 bis 1879 im deutschen buchhandel erschienen sind, nebst angabe einiger ältern werke. Gracklauer's fachkatalog (no.6): Leipzig 1879. pp.18. [125.]

KATALOG der in den jahren 1820–1891 in deutscher sprache erschienenen bücher über fischerei, fischzucht, fischrecht etc. Allgemeine fischereiausstellung: Nürnberg 1892. pp.29. [435.]

Great Britain

INDEXES to the subject matters of the reports of the House of commons. . . . Fisheries, 1803–1833. 1834. pp.[vi].54. [2000.]

[JOHN RUSSELL SMITH], A bibliographical catalogue of english writers on angling and ichthyology. 1856. pp.47. [550.]

also issued as an appendix to Robert Blakey, Historical sketches of the angling literature of all nations [*1856*].

OSMUND LAMBERT, Angling literature in England and descriptions of fishing by the ancients. With a notice of some books on other piscatorial subjects. 1881. pp.[ix].87. [150.]

JAMES ROBB, Notable angling literature. [1946]. pp.229. [500.]

JOHN MOORE, Fishing. National book league: Reader's guide: 1948. pp.20. [75.]
the bibliography proper is by W. A. Munford.

Italy

GUELFO CAVANNA, Elementi per una bibliografia italiana intorno all'idrofauna, agli allevamenti degli animali acquatici e alla pesca. Firenze 1880. pp.viii.170. [1500.]

Japan

EDWARD NORBECK and KATSUNORI SAKURADA, A romanized bibliography of publications in japanese on japanese fishing communities. [Salt Lake City] 1954. ff.[ii].23. [227.]*

New Zealand

DIANA BASSETT, A bibliography of the freshwater fishes of New Zealand. Library school: Bibliographic series (no.5): Wellington 1961. pp. iv.46.

Fish

Norway

THORVALD [OLAF] BOECK, Chronologisk fortegnelse over skrifte og af handlinger om de norske
fiskerier. Christiania 1880. pp.[ii].26. [300.]

FISKERLITTERATUR, Fiskeridirektorat: Bergen
1951 &c.

Russia

V[LADIMIR] K[ONSTANTINOVICH] ESIPOV, Указатель литературы по рыбному хозяйству
европейского севера СССР — 1917–1933 гг.
Арктический институт: Ленинград 1935.
pp.158. [1443.]

N. S. ROMANOV, Указатель литературы
по рыбному хозяйству южных бассейнов
СССР за 1918–1953 гг. Академия наук
СССР: Отделение биологических наук: Ихтиологическая комиссия: Москва 1955.
pp.294. [3906.]

[ELENA SERGEEVNA ZAKHAROVA], Спортив-
ная рыбная ловля. Рекомендательный список литературы. Публичная библиотека
имени М. Е. Салтыкова-Щедрина: Ленинград 1956. pp.12. [40.]

Sweden

FREDRIK LUNBERG, Bidrag till öfversigt af

Fish

Sveriges ichthyologiska literatur. Stockholm
1872. pp.xviii.56. [400.]

Switzerland

FISCHEREI. Centralkommission für schweizerische landeskunde: Bibliographie der schweizerischen landeskunde (section v. 9c.[ii]): Bern 1898.
pp.viii.57. [1000.]
— Supplement . . . (v. 9c.[ii]²): 1916. pp.ix.35.
[600.]

H. FISCHER-SIGWART, Fische. Centralkommission
für schweizerische landeskunde: Bibliographie der
schweizerischen landeskunde (section IV. 6. 5δ):
Bern 1900. pp.xiv.85. [1750.]

Ukraine

D[MYTRO] BELING and O. BILK, Риби прiс-
них вод УСРР. (Бiблiографiчний покажчик).
Українська академія наук: Библіотека:
Серія бібліографічна: Київ 1936. pp.76. [391.]

United States

THEODORE [NICHOLAS] GILL, Bibliography of
the fishes of the Pacific coast of the United States
to the end of 1879. Smithsonian institution: Miscellaneous collections (vol.xxiii = Bulletin of the

United States national museum, no.11): Washington 1882. pp.[iii].73. [750.]

CHA[RLE]S W. SMILEY, List of papers relating to the work of the United States fish commission ... which have been published. United States fish commission: Bulletin: [Washington] 1883. pp. 122. [1309.]
includes the publications of the states.

T. H. MCKEE, Reports of the Committee on fisheries, United States Senate, from the organization of the committee, January 21, 1884, to the close of the forty-ninth congress, 1887. Washington 1887. pp.5. [12.]

LIST of publications of the Bureau of fisheries available for distribution. Bureau of fisheries: Document (no.614): Washington 1907. pp.[ii].20. [300.]
revised at various intervals, the last edition (1934) being a small leaf.

FISHES. Superintendent of documents: Price list (no.21): Washington [1908]. pp.18. [125.]
frequently re-edited under various titles.

ROSE M[ORTIMER] E[LLZEY] MAC DONALD, An analytical subject bibliography of the publications of the Bureau of fisheries, 1871–1920. Bureau of fisheries: Report (1920, appendix v: document 899): Washington 1921. pp.306. [7500.]

[HOWARD H. PECKHAM, HELEN T. GAIGE and CARL L. HUBBS], Ichthyologia et herpetologia americana. A guide to an exhibition in the William L. Clements library illustrating the development of knowledge of american fishes, amphibians and reptiles. William L. Clements library: Bulletin (no.xxv): Ann Arbor 1936. pp.[ii].22. [46.]

A BIBLIOGRAPHY of Iowa fishes. State conservation commission: [Des Moines] 1940. ff.[iii].210 [*sic*, 211].8. [1000.]*

AN AUTHOR bibliography pertaining to Iowa fishes. State conservation commission: [Des Moines] 1940. ff.[iii].114.24.10.3.5. [1000.]*

CHARLES M[CKINLEY] WETZEL, American fishing books. A bibliography from the earliest times up to 1948. Newark, Del. 1950. pp.235. [1250.]
200 copies printed.

ROMEO MANSUETI, A partial bibliography of fish eggs, larvae, and juveniles, with particular reference to migratory and estuarine species of the Atlantic coast and supplemented by a check list and references to the early development of the fishes and fish-like chordates of Maryland water.

Chespeake biological laboratory: [Solomons, Md. 1954]. ff.iii.55. [1496.]*

ROBERT M. JENKINS, Oklahoma fishery publications. A partial bibliography. Oklahoma fishery research laboratory: Norman 1956. pp.28. [300.]*

Yugoslavia

D[INKO] MOROVIĆ, Prilog bibliografiji jadranskog ribarstva. Institut za oceanografiju i ribarstvo u Splitu: Posebna izdanja (vol.i): Split 1950. pp. x.144. [2000.]

3. *Miscellaneous*

[SYLVAIN JOURDAIN], Travaux d'erpétologie et d'ichthyologie de S. Jourdain. Montpellier 1872. pp.10. [17.]

[HENRI ÉMILE SAUVAGE], Notice sur les travaux ichthyologiques de . . . H.-É. Sauvage. Boulogne-sur-Mer [printed] 1893. pp.18. [105.]

J. ARTHUR HUTTON, The literature of fish scales. Manchester 1921. pp.15. [200.]

L[UDWIG] FREUND, Bibliographia pathologiae piscium. Collegit atque edidit auxilio Ministerii pro agricultura Czechoslovakiae [&c.]. Burg bz. Mgdbg [printed] 1923. pp.[ii].189–263. [1250.]

Fish

O[ENE] POSTHUMUS, Otolithi piscium. Fossilium catalogus (I. 24): Berlin 1924. pp.42. [750.]

W. DEECKE, Pisces triadici. Fossilium catalogus (I. 33): Berlin 1926. pp.[ii].201. [4500.]

FISH oils. Science library: Bibliographical series (no.214): 1936. ff.3. [46.]*

MISCELLANEOUS references on insecticides and fish-poisons of vegetable origin. Science library: Bibliographical series (no.317): 1937. ff.14. [257.]*

THE SMOKE-CURING of meat and fish products. Science library: Bibliographical series (no.427): 1938. ff.6. [120.]*

PAUL NEMENYI, An annotated bibliography of fishways, covering also related aspects of fish migration, fish protection and water utilization. University of Iowa studies in engineering: Bulletin (no.23): Iowa City 1941. pp.72. [175.]

[CARL OTTO VON KIENBUSCH], Fresh-water angling. . . . An exhibition. Princeton university: library: [1946]. pp.24. [50.]

CLAUS NISSEN, Schöne fischbücher. . . . Bibliographie fischkundlicher abbildungswerke. Stuttgart 1951. pp.110. [135.]

Fish

WILLIAM MARCUS INGRAM and PETER DOUDOROFF, Publications on industrial wastes relating to fish and oysters. A selected bibliography. Public health service: Publication (no.270 = Public health bibliography series, no.10): Washington 1953. pp. ii.28. [114.]

ROBERT M. JENKINS, *ed.* Bibliography of theses on fishery biology . . . and related subjects. Sport fishing institute: [Washington] 1959. pp.[iii].80. [1743.]★

Natural History.

Natural History

1. Periodicals

CATALOGUE of the natural history library of the Linnean society of London. Part II. Transactions of societies, journals. 1867. pp.[ii].85. [1000.]

C. EKAMA, Catalogue de la bibliothèque. . . . Première livraison: encyclopédies; publications académiques et recueils périodiques. Fondation Teyler: Harlem 1885. pp.[iii].87. [500.]

JENÖ DADAY, A Budapesti könyvtárakban található természetrajzi folyóiratok jegyzéke. Budapest 1890. pp.57. [680.]

CARL SCHMIDT, Synchronistische tabellen über die naturwissenschaftliche journalliteratur von 1650–1893. Schriften herausgegeben von der Naturforscher-gesellschaft bei der universität Jurjew (Dorpat) (vol.viii): Dorpat 1895. pp.x.63. [500.]

AUGUST BÖHM VON BÖHMERSHEIM, Zeitschriftenkatalog des K. k. naturhistorischen hofmuseums. Wien 1897. pp.ix.184. [1566.]
also issued as a supplement to vol.xii of the Hofmuseum's Annalen.

WERNER SPALTEHOLZ, Verzeichnis der periodischen schriften medizinischen und naturwissenschaftlichen inhaltes in der bibliothek der medizinischen und naturwissenschaftlichen institute der universität Leipzig. Dritte auflage. . . . Fortgeführt und erweitert von E. Riecke. Leipzig 1907. pp.104. [1275.]

LISTE des périodiques. Muséum national d'histoire naturelle: Bibliothèque: 1910. pp.[vi].54. [1000.]

[E. MAYR and W. MEISE], Verzeichnis der zeitschriften, die im gebäude des Museums für naturkunde vorhanden sind. Zoologisches museum: Mitteilungen (vol.xiv, supplement): Berlin 1929. pp.viii.188. [1756.]

— — Nachtrag. . . . (vol.xxii, supplement): 1937. ff.[i].34. [478.]

[BASIL H. SOULSBY and A. C. TOWNSEND], Place-numbers of the societies and other corporate bodies issuing serial publications, and of the independent periodical publications. . . . [Second edition]. British museum (natural history): 1930. pp.[vii].175. [2000.]

[JESSEN], Verzeichnis der von der Staats-und universitäts-bibliothek und den instituten der universität gehaltenen zeitschriften aus den gebieten der medizin und naturwissenschaften. Breslau 1931. pp.vii.125. [natural history: 1250.]

LIJST van natuurwetenschappelijke tijdschriften aanwezig in utrechtsche bibliotheken. Utrecht 1931. pp.152. [2000.]

N. WING EASTON, Inventaris van technische en natuurwetenschappelijke periodiken in neder-landsche bibliotheken. 's Gravenhage 1935. pp.ix. 616. [5026.]

PUBLICAÇÕES periódicas estrangeiras inventa-riadas nas bibliotecas portuguesas. Etnologia, ciências naturais, agro-pecuária. Instituto de alta cultura: Centro documentação científica: Lisboa 1953. pp.739. [8000.]

MARGARET HANSELMAN UNDERWOOD, Bibliogra-phy of north american minor natural history serials in the university of Michigan libraries. Ann Arbor 1954. pp.[ix].197. [400.]*

2. *Manuscripts*

A[UGUST] G. [WILHELM] E[DUARD] TH[EODOR] HENSCHEL, Catalogus codicum medii aevi medi-corum ac physicorum qui manuscripti in biblio-thecis vratislaviensibus asservantur. Vratislaviae [1847]. pp.56. [115.]

[J. A.] CLAUDIUS ROUX, Bibliographie métho-dique des principaux manuscrits français relatifs aux sciences naturelles. Société d'agriculture, sciences et industrie: Lyon 1908. pp.112. [1015.]

SPENCER SAVAGE, Calendar of the Ellis manuscripts. The correspondence and miscellaneous papers of John Ellis. Linnean society: Catalogue of the manuscripts in the library (part 4): 1948. pp.vi.104. [300.]

3. *History*

[JOHANN] LUDWIG CHOULANT, Die anfänge wissenschaftlicher naturgeschichte und naturhistorischer abbildung im christlichen abendlande. Ihrem hochverdienten Collegen ... Ernst August Pech ... widmen ... diese denkschrift ... die professoren der Chirurgisch-medicinischen akademie: Dresden 1856. pp.[iv].46. [50.]

[JOHANN] LUDWIG CHOULANT, Graphische incunabeln für naturgeschichte und medicin. Enthaltend geschichte und bibliographie der ersten natur-historischen und medicinischen drucke des XV. und XVI. jahrhunderts, welche mit illustrirenden abbildungen versehen sind. Leipzig 1858. pp.xx.168. [100.]

a facsimile was published in Munich in 1924.

[BERNARD BARHAM WOODWARD], Guide to an exhibition of old natural history books illustrating the origin and progress of the study of natural

history up to the time of Linnæus. British museum
(natural history): Special guides (no.1): 1905.
pp.28. [59.]

FRANCIS J. GRIFFIN, C[HARLES] DAVIES SHERBORN
and H. S. MARSHALL, A catalogue of papers concerning the dates of publication of natural history
books. Society for the bibliography of natural
history: Journal (vol.i, part 1): 1936. pp.30. [600.]

F[ERDINANDO] M[ARIA] SCAPIN, Catalogo degli
incunaboli medici e naturalistici della biblioteca
universitaria di Padova. [Padova] 1956. pp.69.
[101.]

4. *General*

JOHANN JACOB SCHEUCHZER, Bibliotheca scriptorum historiæ naturalis omnium terræ regionum
inservientium. . . . Accessit . . . Jacobi Le Long . . .
de scriptoribus historiæ naturalis Galliæ. Tiguri
1716. pp.[xiii].241. [1250.]
reissued with the date 1751.

FRIEDRICH BOERNER, Bibliothecae librorvm rario-
rvm physico-medicorvm historico-criticae specimen primvm[–secvndvm]. Helmstadii [1751–]
1752. pp.[viii].36+[viii].37–67. [32.]

JULIUS BERNHARD VON ROHR, Physikalische
bibliothek, worinnen die vornehmsten schriften

die zur naturlehre gehören, angezeiget werden,
... herausgegeben von Abraham Gotthelf Kästner.
Leipzig 1754. pp.[xxxii].964.[xxx]. [3000.]
a copy in the British museum contains ms. additions.

FRIEDRICH BOERNER, Relationes de libris physico-
medicis partim antiqvis partim raris. . . . Fasci-
cvlvs I. Vitembergae 1756. pp.[x].150. [30.]
no more published.

JOHANN CARL HEFFTER, Museum disputatorium
physico-medicum tripartitum. Zittaviae Lvsa-
torvm 1756–1764. pp.[xl].480.80.84+14.298+
[ii].299–526.92.88. [18,498.]
*the first volume was reissued in 1763 as a so-called
'Editio nova'.*

LAURENZ THEODOR GRONOVIUS, Bibliotheca
regni animalis atque lapidei, seu recensio auctorum et librorum, qui de regno animali & lapideo
... tractant. Lugduni Batavorum 1760. pp.[viii].
326. [5000.]

[BARON OTTO VON MÜNCHHAUSEN], Des haus-
vaters zweyten theils erstes stuck. Inhalt: des haus-
vaters botanische, physikalische und oekonomische bibliothek. Hannover 1765–1766. pp.[ii].
xxxvi.[xii].367+369–832.[ciii]. [4000.]

E[RNST] G[OTTFRIED] BALDINGER, Biographien

jetzlebender aerzte und naturforscher in und
ausser Deutschland. Jena [1768–]1772. pp.[xxiv].
120 + [xiv].123–264 + [xiv].129 + [vi].129–250.
[1500.]

JOHANN BECKMANN, Physikalisch-ökonomische
bibliothek, worinn von den neuesten büchern,
welche die naturgeschichte, naturlehre und die
land- und stadtwirthschaft betreffen, zuverlässige
und vollständige nachrichten ertheilet werden.
Göttingen.

    i. 1770. pp.[x].654.[xxvi]. [85.]
    ii. 1771. pp.[iii].626.[xxiii]. [80.]
    iii. 1772. pp.[vi].613.[xxv]. [112.]
    iv. [1773–]1774. pp.[vi].616.25. [114.]
    v. 1774. pp.[vi].616.[xxiii]. [120.]
    vi. 1775. pp.[vi].600.[xxiii]. [130.]
    vii. 1776. pp.[viii].620.[xxiv]. [139.]
    viii. 1777. pp.[viii].616.[xxiv]. [134.]
    ix. 1778. pp.[viii].608.[xxii]. [124.]
    x. 1779. pp.[viii].604.[xxiii]. [132.]
    xi. [1780–]1781. pp.[viii].598.[xx]. [166.]
    xii. [1782–]1783. pp.[x].614.[xx]. [179.]
    xiii. [1784–]1785.pp.[viii].598.[xxii]. [196.]
    xiv. [1785–]1787, pp.[vi].616.[xxiii]. [169.]
    xv. [1787–]1789. pp.[viii].604.[xxiv]. [162.]

xvi. [1789–]1791. pp.[vi].598.[xx]. [143.]
xvii. [1791–]1793. pp.[vi].602.[xxii]. [142.]
xviii. [1793–]1795. pp.[viii].620.[xxiv]. [145.]
xix. [1795–]1797. pp.[viii].598.[xxiii]. [138.]
xx. 1798. pp.[vi].620.[xxvii]. [121.]
xxi. [1800–]1802. pp.[x].602.[xxx]. [145.]
xxii. [1803–]1804. pp.[x].608.[xxxviii]. [141.]
xxiii. [1805–]1806. pp.[viii].586. [121.]

JOHANN TRAUGOTT MÜLLER, Einleitung in die
oekonomische und physikalische bücherkunde
und in die damit verbundenen wissenschaften bis
auf die neuesten zeiten. Leipzig 1780–1784. pp.
[xvi].560+[ii].viii.718+[iv].iv.743. [15,000.]

J[OSEPH] P[AUL] VON COBRES, Deliciae cobre-
sianae. . . . Büchersammlung zur naturgeschichte.
Augsburg [printed] 1782. pp.[iii].xxviii.470+[ii].
471–957. [2500.]

JOHANN SAMUEL SCHRÖTER, Für die litteratur
und kenntniss der naturgeschichte, sonderlich der
conchylien und der steine. Weimar 1782. pp.
[xiv].278.[viii]+[xii].314.[viii]. [125.]

[C. F. PRANGE], Systematisches verzeichniss aller
derjenigen schriften welche die naturgeschichte
betreffen; von den ältesten bis auf die neuesten
zeiten. Halle 1784. pp.viii.446. [2500.]

JOHANN SAMUEL SCHRÖTER, Neue litteratur und beyträge zur kenntniss der naturgeschichte, vorzüglich der conchylien und fossilien. Leipzig 1784–1787. pp.[viii].550.[xxx] + [viii].598.[xxxiv] + [vi.614.[xxvii]+[xii].456.[xxvi]. [500.]

GEORG RUDOLPH BOEHMER, Bibliotheca scriptorum historiae naturalis, oeconomiae aliarumque artium ac scientiarum ad illam pertinentium realis systematica. Lipsiae 1785–1789. pp.[ii].xviii.778+ [iii].772 + [iii].604 + [ii].xxx.536 + 808 + 642 + [ii].510+[ii].412+[iii].x.3–740. [65,000.]

DESIDERATA pro Bibliotheca banksiana. [London] 1790. pp.27. [1343.]

[JOHANN SAMUEL ERSCH], Allgemeines repertorium der literatur für die jahre 1785 bis 1790. x. Physikalisch-naturhistorische literatur. Jena 1793. pp.[67]. [1729.]
[—] — 1791–1795. Weimar 1800. pp.[86]. [1788.]
[—] — 1796–1800. Weimar 1807. pp.[72]. [1078.]
no more published.

[JOHANN SAMUEL ERSCH], Systematisches verzeichniss der in der medicinischen, physicalischen, chemischen und naturhistorischen literatur in den

jahren 1785 bis 1790 herausgekommenen deutschen und ausländischen schriften. Jena 1795. pp.
[159]. [natural history: 2000.]

—— 1791–1795. Weimar 1799. pp.[236].
[4000.]

JOHANN DRYANDER, Catalogus bibliothecæ
historico-naturalis Josephi Banks. Londini 1798–
1800. pp.vii.309.[xii]+xx.578.[xxx]+xxiii.656.
[xxxviii]+ix.390.[xxvi]+[ii].514. [25,000.]
*this collection now forms part of the British museum
library.*

1. [JEREMIAS DAVID] REUSS, Repertorium commentationum a societatibus litterariis editarum.
. . . Tom. 1. Historia naturalis, generalis et zoologia. Gottingae 1801. pp.[iii].iv.74.xx.75–574.
[natural history: 600.]

KARL FRIEDRICH BURDACH, Handbuch der neuesten in- und ausländischen literatur der gesammten naturwissenschaften, und der medicin und
chirurgie. Gotha 1821. pp.[ii].xiv.392. [4374.]
also forms vol.iii of (= 1st supplement to) Burdach's
Die literatur der heilwissenschaft, *1810–1811, which
is entered under Medicine, above.*

REVUE bibliographique pour servir de com-
plément aux Annales naturelles; par [Jean Victor]

Audouin, Ad[olphe Théodore] Brongniart et [Jean Baptiste] Dumas.

    1829. pp.[iii].156. [529.]
    1830. pp.[iii].144. [228.]
    1831. pp.96. [200.]

no more published; the issue for 1831 is incorrectly described as the '2e année'.

[JOHN KIDD], Catalogue of the works in medicine and natural history contained in the Radcliffe library. Oxford 1835. pp.viii.330. [6000.]

CATALOGUE of the library of the Boston society of natural history. Boston 1837. pp.27. [503.]

WILHELM ENGELMANN, Bibliotheca historico-naturalis. Verzeichniss der bücher . . . welche in Deutschland, Scandinavien, Holland, England, Frankreich, Italien und Spanien in den jahren 1700–1846 erschienen sind. . . . Erster band. Bücherkunde, hülfsmittel, allgemeine schriften, vergleichende anatomie und physiologie, zoologie, palaeontologie. Leipzig 1846. pp.ix.786. [10,000.]

no more published; reissued in 1847 with an english titlepage, Literature of natural history.

— — Supplement-band, enthaltend die in den periodischen werken aufgenommenen und die vom jahre 1846–1860 erschienenen schriften. . . .

Von J. Victor Carus . . . und Wilhelm Engelmann.
1861. pp.x.950+xxiv.951–2144. [40,000.]
   *a continuation of this work is entered under Zoology,
below.*

NATURWISSENSCHAFTLICHES literaturblatt. Bei-
lage zur 'Natur'. Halle.
   1856. pp.24. [30.]
   1857. pp.96. [90.]
   1858. pp.64. [50.]
   1859. pp.64. [50.]
   1860. pp.32. [30.]
   1861. pp.32. [25.]
   1862. pp.32. [50.]
   1863. pp.32. [40.]
   1864. pp.32. [40.]
   1865. pp.24. [30.]
   1866. pp.16. [20.]
   1867. pp.24. [30.]
   1868. pp.12. [20.]
   *no more published; issued as supplements to vols.v–
xvii of* Die natur.

CATALOGUE of the natural history library of
the Linnean society. 1866–1867. pp.[iii].289+
[ii].85. [9000.]
   — Part III. Additions. 1877. pp.[ii].109. [2000.]

P. PUZUIREVSKI, Каталогъ библіотеки Им-

ператорскаго С.-Петербургскаго минерало-
гическаго общества. С.-Петербургъ 1867. pp.
viii.234. [natural history: 2000.]

CATALOGO della biblioteca. Società italiana di
scienze naturali: Milano 1868. pp.47. [1000.]

ÁGOST HELLER, A Kir. magyar természettudományi társulat könyveinek czímjegyzéke. Budapest 1877. pp.332. [4528.]

[—] — [another edition]. Összeállította Ráth
Arnold. 1901. pp.[ix].coll.566. [11,240.]

NATURAE novitates. Bibliographie neuer er-
scheinungen aller länder auf dem gebiete der
naturgeschichte und der exacten wissenschaften.
Herausgegeben von R. Friedländer & sohn [vols.
lxiv–lxv: Paul Budy vorm. R. Friedländer &
sohn]. Berlin.

    i. 1879. pp.[ii].276. [5000.]
    ii. 1880. pp.[ii].238. [5000.]
    iii. 1881. pp.[ii].246. [5000.]
    iv. 1882. pp.[iii].276. [6000.]
    v. 1883. pp.[iii].284. [6000.]
    vi. 1884. pp.[ii].298. [6000.]
    vii. 1885. pp.[ii].320. [6000.]
    viii. 1886. pp.[ii].334. [6000.]
    ix. 1887. pp.[ii].386. [6500.]
    x. 1888. pp.[ii].466. [7000.]

xi. 1889. pp.[ii].466. [7000.]
xii. 1890. pp.[ii].570. [6780.]
xiii. 1891. pp.[ii].562. [6714.]
xiv. 1892. pp.[ii].526. [6955.]
xv. 1893. pp.[ii].600. [8149.]
xvi. 1894. pp.[ii].662. [9140.]
xvii. 1895. pp.[ii].806. [8286.]
xviii. 1896. pp.[ii].684. [9116.]
xix. 1897. pp.[ii].686. [9024.]
xx. 1898. pp.[ii].780. [9359.]
xxi. 1899. pp.[ii].866. [9430.]
xxii. 1900. pp.[ii].670. [7786.]
xxiii. 1901. pp.[ii].734. [8608.]
xxiv. 1902. pp.[ii].780. [8766.]
xxv. 1903. pp.[ii].752. [9320.]
xxvi. 1904. pp.[ii].772. [9585.]
xxvii. 1905. pp.[ii].688. [8523.]
xxviii. 1906. pp.[ii].722. [9334.]
xxix. 1907. pp.[ii].684. [9017.]
xxx. 1908. pp.[ii].704. [9155.]
xxxi. 1909. pp.[ii].660. [8720.]
xxxii. 1910. pp.[ii].648. [8272.]
xxxiii. 1911. pp.[ii].652. [8344.]
xxxiv. 1912. pp.[ii].712. [9492.]
xxxv. 1913. pp.[ii].676. [8422.]
xxxvi. 1914. pp.[ii].542. [7459.]
xxxvii. 1915. pp.[ii].378. [5044.]

xxxviii. 1916. pp.[ii].322. [4168.]
xxxix–xl. 1916–1917. pp.[ii].482. [6066.]
xli. 1919. pp.[ii].222. [2953.]
xlii. 1920. pp.[ii].286. [3765.]
xliii. 1921. pp.[ii].314. [3781.]
xliv. 1922. pp.[ii].298. [3877.]
xlv. 1923. pp.[ii].246. [3749.]
xlvi. 1924. pp.[ii].226. [3687.]
xlvii. 1925. pp.[ii].226. [3730.]
xlviii. 1926. pp.[ii].224. [3694.]
xlix. 1927. pp.[ii].224. [3647.]
l. 1928. pp.[ii].222. [3497.]
li. 1929. pp.[ii].212. [3896.]
lii. 1930. pp.[ii].214. [3300.]
liii. 1931. pp.[ii].214. [3389.]
liv. 1932. pp.[ii].198. [3310.]
lv. 1933. pp.[ii].214. [3444.]
lvi. 1934. pp.[ii].206. [3155.]
lvii. 1935. pp.[ii].198. [3222.]
lviii. 1936. pp.[ii].216. [3460.]
lix. 1937. pp.[ii].198. [3165.]
lx. 1938. pp.[ii].198. [3185.]
lxi. 1939. pp.[ii].160. [2844.]
lxii. 1940. pp.[ii].112. [1719.]
lxiii. 1941. pp.96. [1836.]
lxiv. 1942. pp.96. [1796.]
lxv. 1943. pp.112. [1051.]
no more published.

WOLF'S naturwissenschaftlich-mathematisches vademecum. [Leipzig 1881]. pp.[ii].250. [5000.]

CATALOGUE of the books in the library of the Bristol naturalists' society. Clifton [printed] 1881. pp.[ii].24. [350.]

CATALOGUE of books in the library of the natural history society. Glasgow 1883. pp.20. [400.]

JOHN HOPKINSON, Catalogue of the library of the Hertfordshire natural history society and field club. 1885. pp.52. [1250.]

C. EKAMA [vol.iii: G. C. W. BOHNENSIEG], Catalogue de la bibliothèque. . . . Tome I [III]. Sciences exactes et naturelles. Fondation Teyler: Harlem 1885–1888, 1904. pp.vii.827+vii.1168.2. [13,500.]
vol.ii is entered under other headings.

CATALOGUE of the Colonial museum library. New Zealand [printed Wellington] 1890. pp.[iii]. 56. [1000.]

LIST of the [current] natural history publications of the trustees of the British museum. [1894]. pp.24. [350.]
— [another edition]. [1954]. pp.32. [600.]
there are numerous intermediate editions.

CATALOGUE de la bibliothèque. 1ᵉʳ [II] fascicule. Société linnéenne: Bordeaux 1894, 1901. pp.[iii]. 175+[v].115. [5000.]

WILLIAM E. HOYLE, A catalogue of the books and pamphlets in the library. Owens college: Manchester museum: Manchester 1895. pp.xvi. 302. [4000.]

CATALOGUE of books in the library of the Brighton & Sussex natural history and philosophical society. Brighton 1895. pp.[iv].41. [1000.]

HARRIET HOWARD STANLEY, Reading list of out-of-door books. University of the state of New York: State library bulletin (Bibliography no.8): [Albany] 1898. pp.157–177. [125.]

RANDOLPH I[LTYD] GEARE, A list of the publications of the United States National museum (1875–1900). United States National museum Bulletin (no.51): Washington 1902. pp.vi.168. [1300.]

— — Supplement. 1906. pp.40. [300.]

[—] — [another edition]. A list and index of the publications [&c.] . . . (no.193): 1947. pp.iii. 306. [4000.]

[BERNARD BARHAM WOODWARD], Catalogue of the books, manuscripts, maps and drawings in the British museum (natural history). 1903–1915.

pp.viii.500 + [v].501–1038 + [v].1039–1494 + [v].1495–1956+[v].1957–2403. [90,000.]

— — Addenda and corrigenda. [1922]. pp. [ii].48.

[—] — Supplement. 1922–1940. pp.[v].511+ [v].513–967+[iv].969–1480. [30,000.]

AANWINSTEN op het gebied der wis- en natuur-kundige wetenschappen, gedeeltelijk behoorende tot . . . C. P. Burger. Universiteit: Bibliotheek: Amsterdam 1907. pp.[ii].32. [600.]

THE RUSSELL [Gurdon Wadsworth Russell] collection of books on natural history in Trinity college library. Trinity college: Bulletin (n.s. x, no.2): Hartford 1913. pp.28. [125.]

CATALOGUS van de boeken en tijdschriften geplaatst in de wiskundige leeszaal en in de Laboratoria voor natuurkunde, dierkunde, plant-kunde, plantenphysiologie en pharmacognosie. Universiteit: Bibliotheek: Amsterdam 1922. pp. 95. [1500.]

CATALOGUE of books in the library at the Athenæum, Bury St. Edmunds. Suffolk institute of archæology and natural history: [Bury St. Edmunds] 1933. pp.30. [750.]

BRENT ALTSHELER, Natural history index-guide. Index to the books in libraries. Louisville, Ky.

1936. pp.311. [35,000.]
—— Second edition. 1940. pp.583. [45,000.]

SCIENTIAE naturalis bibliographia. Ed.: W[il-helm] Junk. Annus I, Pars I. Den Haag 1937. pp.29. [600.]
no more published.

RICHARD JAMES HURLEY, Key to the out-of-doors. A bibliography of nature books and materials. New York 1938. pp.256. [3000.]

CATALOGUE of the library. South-eastern union of scientific societies: [1938]. pp.27. [300.]

HERBERT FAULKNER WEST, The nature writers. A guide to richer reading. Brattleboro 1939. pp. 155. [300.]

F[RANÇOIS] BOURLIÈRE, Éléments d'un guide bibliographique du naturaliste. Mâcon 1940. pp. ix.303. [5813.]
—— Supplément I et II. 1941. pp.303–369. [550.]

HUGH PILCHER, Natural history. National book council: Book list (no.182): 1942. pp.[4]. [140.]

LISTE des acquisitions. Muséum national d'histoire naturelle: Bibliothèque centrale.★
1948. pp.33. [400.]
1949. pp.29. [350.]

1950. ff.34. [400.]
1951. ff.48. [600.]
1952. ff.51. [600.]
1953. ff.42. [500.]
1954. ff.66. [800.]
1955. ff.84. [1000.]
1956. ff.71. [800.]
1957. ff.46. [500.]
1958. ff.71. [800.]
1959. ff.66. [700.]
1960. ff.65. [700.]
1961. ff.94. [1000.]

in progress.

NATURWISSENSCHAFTEN, medizin. Ein verzeichnis von büchern u. aufsätzen. Stadtbibliothek: [Hanover] 1954. pp.v.146. [6000.]*

SUZANNE LAVAUD, Catalogue des thèses de doctorat ès sciences naturelles soutenues à Paris de 1891 à 1954. Faculté de pharmacie: Bibliothèque: Paris 1955. pp.257. [1181.]

LIST of accessions to the museum library. British museum (natural history).

1955. pp.301.97. [2196.]
1956. pp.241.77. [1729.]
1957. pp.203. [1357.]
1958. pp.192. [1533.]

1959. pp.168.29. [1248.]
1960. pp.218.37. [1528.]
1961. pp.332.50. [2139.]
1962.
1963. pp.457. [2687.]
in progress.

HELENA WINKLEROVÁ, Knižky o přírodě. Universita: Knihovna: Čteme a studujeme (1956, no.5): [Prague 1956]. pp.32. [200.]

VERÖFFENTLICHUNGEN aus dem Zoologischen institut der Technischen hochschule und dem Staatlichen naturhistorischen museum zu Braunschweig in den jahren 1947 bis 1955. Braunschweig 1956. ff.[i].22. [150.]*

HELENA WINKLEROVÁ, Co číst o živé přírodě. Universita: Knihovna: Čteme a studujeme (1957, no.10): Praze [1957]. pp.32. [300.]

ILUŠE CEJPKOVÁ, JARMILA BURGETOVÁ and HELENA WINKLEROVÁ, Poznáváme přírodu. (Sborník zkušeností z propagace přírodovědecké literatury mezi dětmi a mládeží v lidových knihovnách). Universita: Knihovna: Čteme a studujeme (1958, no.5): Praze [1958]. pp.76. [550.]

NATIONAL nature week. . . . A list of books on all aspects of natural history. Public library: [Rotherham 1963]. pp.28. [600.]*

214

Natural History

5. Countries

America

CATALOGUE of the north american natural history library of John Lewis Childs. Floral Park, N.Y. 1917. pp.[v].150. [2250.]

MAX MEISEL, A bibliography of american natural history. The pioneer century, 1769–1865. Brooklyn, N.Y. 1924–1929. pp.244+xii.741+ xii.749. [12,500.]

Australia

S. SINCLAIR, Catalogue of the library of the Australian museum. [Second edition]. Part III. — Pamphlets. Sydney 1893, 1905. pp.[iv].27+[ii]. 29–84. [1500.]
no more published.

Brazil

ACHEGAS para a bibliographia das sciencias naturaes. Resumo de obras, opusculos e artigos publicados no estrangeiro e interessando o Brasil (1917–1921). São Paulo 1927. pp.[ii].210. [500.]

Denmark and Norway

M[ORTEN] TH[RANE] BRÜNNICH, Literatura danica scientiarum naturalium, qva comprehenduntur I. Les progrès de l'histoire naturelle en Dannemarc

& en Norvège. II. Bibliotheca patria auctorum & scriptorum, scientias naturales tractantium. Hafniæ &c. 1783. pp.[ii].124.242.xiv.[xvi]. [2000.]

M. WINTHER, Literaturæ scientiæ rerum naturalium in Dania, Norvegia & Holsatia usque ad annum MDCCCXXIX. Havniæ 1829. pp.xvii.233. [xv]. [2000.]

France

LOUIS ANTOINE PROSPER HÉRISSANT, Bibliothèque physique de la France, ou liste de tous les ouvrages, tant imprimés que manuscrits, qui traitent de l'histoire naturelle de ce royaume. . . . Ouvrage achevé & publié par M. ★★★ [C. J. L. Coquereau]. 1771. pp.40.496. [1650.]

[J. A.] CLAUDIUS ROUX, Notice bibliographique sur plus de deux cents manuscrits inédits ou peu connus concernant pour la plupart l'histoire naturelle de la région lyonnaise. Société linnéenne de Lyon: Lyon 1906. pp.26. [210.]

[J. A.] CLAUDIUS ROUX, Résumé historique et analytique sur la vie et les travaux des principaux naturalistes foréziens. Lyon 1907–1910. pp.7+52. [410.]

L. TOLMER, Index bibliographique des travaux de sciences naturelles concernant la Normandie, le Maine, l'Anjou et le Blésois pour la décade 1923–1933. Caen [1935]. pp.191. [2329.]

ABEL JEANDET, Recherches bio-bibliographiques pour servir à l'histoire des sciences naturelles en Bourgogne et particulièrement dans le département de Saône-et-Loire depuis le xvi⁰ siècle jusqu'à nos jours. Mâcon [printed] 1892. pp.[v]. 133. [200.]

60 copies privately printed.

[PIERRE DOIGNON], Répertoire bibliographique et analytique du massif de Fontainebleau et de la basse vallée du Loing. Association des naturalistes de la vallée du Loing: La Forêt de Fontainebleau, recherches sur son sol, sa faune, sa flore (no.13): Fontainebleau 1958. pp.56. [3000.]

Germany

FRIEDRICH BÖRNER, Nachrichten von den vornehmsten lebensumständen und schriften jetztlebender berühmter aerzte und naturforscher in und um Deutschland. Wolfenbüttel [1748–]1749–1764.

details of this work are entered under Medicine: Germany, above.

JOHANN SAMUEL ERSCH, Handbuch der deutschen literatur seit der mitte des achtzehnten jahrhunderts. . . . Des zweyten bandes erste . . . abtheilung, die literatur der mathematik, natur- und gewerbskunde enthaltend. Amsterdam &c. 1813. pp.xii.coll.760. [natural history: 2500.]

—— Neue . . . ausgabe, von Franz Wilhelm Schweigger-Seidel. Leipzig 1828. pp.[x].coll. 1740. [6000.]

FR. X. LEHMANN, Die litteratur für vaterlän- dische naturkunde im grossherzogtum Baden. Karlsruhe 1886. pp.44. [575.]

[ADOLF MEYER], Naturforschung und natur- lehre im alten Hamburg. Erinnerungsblätter zu ehren der 90. versammlung der Gesellschaft deut- scher naturforscher und ärzte. Staats- und univer- sitäts-bibliothek: Hamburg 1928. pp.viii.99. [453.]

STEPHAN AUMÜLLER, Allgemeine bibliographie des Burgenlandes. II. teil. Naturwissenschaften. Burgenländisches landesarchiv und Burgenlän- dische landesbibliothek: Eisenstadt 1956. pp.93. [1169.]

RICHARD IMMEL, Das schrifttum über forstwesen, holzwirtschaft, jagd, fischerei und naturschutz in Hessen und Rheinland-Pfalz, unter berücksichti- gung angrenzender gebiete von Baden-Württem- berg, Bayern und Nordrhein-Westfalen. Mainz 1958. pp.[xii].1040. [16,000.]

Natural History

Great Britain

SELECTED list of books on nature study. Public libraries: Sunderland 1911. pp.10. [200.]

F. C. MORGAN, Catalogue of the library of the Woolhope naturalists' field club. Hereford [printed] 1941. pp.92. [600.]

A CATALOGUE of mss., books and periodicals devoted to Gilbert White and natural history. Ealing public libraries: Selborne society library: [Ealing] 1958. pp.34. [200.]*

Hungary

JÓZSEF SZINNYEI and JÓZSEF SZINNYEI, Magyarország természettudományi és mathematikai könyvészete, 1472–1875. Kir. magyar természettudományi társulat: Budapest 1878. pp.viii. coll.1008. [15,000.]

India

CATALOGUE of the library of the Madras government museum. Madras 1894. pp.243. [10,000.]

Italy

J. JOSEPHUS [GIOVANNI GIUSEPPE] BIANCONI, Repertorio italiano per la storia naturale. Repertorium italicum complectens zoologiam, mineralogiam, geologiam et palaeontologiam. Bononiae 1853–1854. pp.vii.192+v.192. [750.]

MARIO CERMENATI, La Valtellina ed i naturalisti. Memoria bibliografica. Sondrio 1887. pp.272. xxxii.273–287. [750.]

PIETRO CAPPARONI, Profili bio-bibliografici di medici e naturalisti celebri italiani dal sec. xv° al sec. xviii°. Istituto nazionale medico farmacologico 'Serono': Roma 1926–1928. pp.117+138. [1500.]

Micronesia

HUZIO UTINOMI [FUJIO UCHINOMI], Bibliographia micronesica scientiae naturalis et cultus. Tokyo 1944. pp.[ii].3.211. [3500.]
—— [another edition]. Bibliography of Micronesia (bibliographia micronesia: scientiae naturalis et cultus). Edited . . . by O. A. Bushnell. Pacific area bibliographies: Honolulu [1952]. pp.xiv.157. [3000.]

Netherlands

ALGEMEENE aardrijkskundige bibliographie van Nederland. . . . Tweede deel. Natuurkundige toestand, bewerkt door W. J. D. van Iterson [*and others*]. Nederlandsch aardrijkskundige genootschap: Leiden 1888. pp.x.249. [4000.]

PALFIJN. Tweemaandelijksch bibliographisch tijdschrift voor nederlandsche natuur- en genees-kundige literatuur. Ghent &c. 1902.
no more published.

CATALOGUS van boeken in Noord-Nederland verschenen van de vroegsten tijd tot op heden. [IX. Wis- en natuurkunde]. Samengesteld door de Tentoonstellings-commissie der Nationale ten-toonstelling van het boek . . . 1910. Vereeniging ter bevordering van de belangen des boekhandels: 's-Gravenhage 1911. pp.[ii].coll.96.[viii]. [1250.]

Russia

[ERNST GOTTFRIED] BALDINGER, Russische phy-sisch-medicinische litteratur dieses jahrhunderts. . . . Erstes stück. Teutsche aerzte und natur-forscher in Russland, von Peter I. bis Catharina II. Marburg 1792. pp.62. [200.]
no more published.

M[ARIYA] M[IKHAILOVNA] TATARINOVA, Ни-зовья Аму-Дарьи. Природные условия и сельское хозяйство. Указатель основной литературы. Центральная научная сельско-хозяйственная библиотека: Ташкент 1961. pp.236. [1901.]

Natural History

Scotland

LIST of selected books on wild life in Scotland. Corporation public libraries: Glasgow 1950. pp. 16. [150.]

Sweden

CARL GUST. WARMHOLTZ, Bibliotheca historica sveo-gothica. . . . Andre delen, som innehåller de böcker och skrifter, hvilka angå Sveriges natural-historia. Stockholm 1783. pp.xxii.141. [ix]. [400.]

Switzerland

[BARON GOTTLIEB EMANUEL VON] HALLER, Catalogue raisonné des auteurs qui ont écrit sur l'histoire naturelle de la Suisse. [Berne] 1777. pp.150. [573.]

VERZEICHNIS der gegenwärtig im archive der Allgemeinen schweizerischen gesellschaft für die gesammten naturwissenschaften sich befindlichen bücher und bildnisse. Aarau 1836. pp.[ii].30. [300.]

— [another edition]. Verzeichnis der im archive der Schweizerischen naturforschenden gesellschaft sich bildenden bibliothek. [By] R. Wolf. Bern 1843. pp.viii.52. [350.]

— [another edition]. Verzeichnis der in der bibliothek der Schweizerischen naturforschenden gesellschaft vorhandenen bücher. [By Christian Christener]. 1850. pp.[iv].172. [2000.]

— [another edition]. Verzeichnis der bibliothek der Schweizerischen naturforschenden gesellschaft. [By J. R. Koch]. 1864. pp.x.188. [4500.]

United States

CAROLUS [CHARLES FRÉDÉRIC] GIRARD, Bibliographia americana historico-naturalis a.d. 1851. Smithsonian institution: Smithsonian report [no.48]: Washington 1852. pp.[ii].iv.60. [281.]

MARY ELLIS, Index to publications of the New York state natural history survey and New York state museum, 1837–1902, also including other New York publications on related subjects. University of the state of New York: Bulletin (no.288 = State museum: Bulletin, no.66): Albany 1903. pp.[ii].239–653. [8000.]

LIST of recent nature writers in America. Library of Congress: Washington 1921. ff.5. [68.]*

WILLIAM MARTIN SMALLWOOD and MABEL SARAH COON SMALLWOOD, Natural history and the american mind. Columbia studies in american culture (no.8): New York 1941. pp.xvii.445. [800.]

PUBLICATIONS, 1941–1950. Wildlife research laboratory: Denver 1952. pp.13. [125.]*

Natural History

JOHN VAN OOSTEN, Great lakes fauna, flora and their environment. A bibliography. Great lakes commission: Ann Arbor 1957. pp.x.86. [2500.]*

DANIEL MCKINLEY, A chronology and bibliography of wildlife in Missouri. University of Missouri: Bulletin (vol.61, no.13 = Library series, no.26): Columbia 1960. pp.128. [1750.]

6. Education

CLARA WHITEHILL HUNT, Illustrative material for nature study in primary schools. University of the state of New York: State library bulletin (Bibliography no.16): [Albany] 1899. pp.463–491. [400.]

WILLIAM GOULD VINAL, Nature education. A selected bibliography. Western reserve university: School of education: Curriculum laboratory (no. 39): Cleveland 1934. ff.82. [1988.]*

7. Miscellaneous

INDEX scriptorum ad medicinam et scientias naturales spectantium d-ris Ludovici Adolphi Neugebaueri. [Warsaw 1870]. pp.15. [70.]

IDA S[OPHIA] SIMONSON, Through the year — days and seasons. Stories and poetry. Revised. Northern Illinois state teachers college quarterly (vol.xxvi, no.1): De Kalb 1930. pp.170. [2750.]

Ornithology.

1. *General*

L[OUIS JEAN RODOLPHE] AGASSIZ, Nomina systematica generum avium. [Soloduri 1842–1846]. pp.xii.90.24. [4000.]

GEORGE ROBERT GRAY, Hand-list of genera and species of birds. [British museum]: 1869–1871. pp.xx.404+xv.278+xi.350. [20,000.]

C[HRISTOPH] G[OTTFRIED ANDREAS] GIEBEL,

225

Ornithology

Thesaurus ornithologiae. Repertorium der gesammten ornithologischen literatur. Leipzig 1872–1877. pp.xi.868+vi.788+vi.861. [50,000.]

THE ZOOLOGICAL RECORD [III: *afterwards:* XVII; XV; XIV; XVI; XVII]. Aves.

xiii. 1876. By Osbert Salvin. pp.60. [750.]

xiv. 1877. By Howard Saunders. pp.59. [750.]

xv. 1878. pp.60. [750.]

xvi. 1879. pp.64. [750.]

xvii. 1880. pp.49. [500.]

xviii. 1881. pp.52. [600.]

xix. 1882. By R. Bowdler Sharpe. pp.48. [600.]

xx. 1883. pp.44. [500.]

xxi. 1884. By A. H. Evans. pp.68. [750.]

xxii. 1885. pp.61. [600.]

xxiii. 1886. pp.69. [500.]

xxiv. 1887. pp.76. [600.]

xxv. 1888. pp.95. [750.]

xxvi. 1889. pp.79. [600.]

xxvii. 1890. By R. Bowdler Sharpe. pp.64. [500.]

xxviii. 1891. pp.69. [500.]

xxix. 1892. pp.63. [400.]

xxx. 1893. pp.49. [400.]

Ornithology

xxxi. 1894. pp.55. [400.]
xxxii. 1895. pp.52. [424.]
xxxiii. 1896. pp.61. [639.]
xxxiv. 1897. pp.54. [567.]
xxxv. 1898. pp.58. [568.]
xxxvi. 1899. pp.72. [714.]
xxxvii. 1900. pp.64. [580.]
xxxviii. 1901. pp.77. [803.]
xxxix. 1902. pp.71. [627.]
xl. 1903. pp.72. [724.]
xli. 1904. pp.71. [679.]
xlii. 1905. pp.88. [742.]
xliii. 1906. pp.104. [1250.]
xliv. 1907. pp.139. [1716.]
xlv. 1908. pp.148. [1949.]
xlvi. 1909. By W. L. Sclater. pp.130. [1721.]
xlvii. 1910. pp.128. [1708.]
xlviii. 1911. pp.115. [1536.]
xlix. 1912. pp.135. [1665.]
l. 1913. pp.127. [1576.]
li. 1914. pp.77. [1088.]
lii. 1915. pp.74. [934.]
liii. 1916. pp.72. [942.]
liv. 1917. pp.62. [707.]
lv. 1918. pp.75. [937.]
lvi. 1919. pp.71. [837.]
lvii. 1920. pp.66. [832.]

lviii. 1921. pp.63. [819.]

lix. 1922. pp.71. [879.]

lx. 1923. pp.75. [1027.]

lxi. 1924. pp.75. [988.]

lxii. 1925. pp.70. [971.]

lxiii. 1926. pp.77. [1089.]

lxiv. 1927. pp.89. [1296.]

lxv. 1928. pp.93. [1319.]

lxvi. 1929. pp.91. [1317.]

lxvii. 1930. pp.97. [1406.]

lxviii. 1931. pp.121. [1764.]

lxix. 1932. pp.101. [1385.]

lxx. 1933. pp.97. [1471.]

lxxi. 1934. pp.106. [1556.]

lxxii. 1935. pp.117. [1762.]

lxxiii. 1936. pp.102. [1570.]

lxxiv. 1937. pp.116. [1699.]

lxxv. 1938. pp.113. [1713.]

lxxvi. 1939. pp.99. [1446.]

lxxvii. 1940. pp.60. [830.]

lxxviii. 1941. pp.82. [1240.]

lxxix. 1942. pp.75. [1076.]

lxxx. 1943. pp.76. [1069.]

lxxxi. 1944. pp.88. [1490.]

lxxxii. 1945. pp.85. [1422.]

lxxxiii. 1946. pp.77. [1265.]

lxxxiv. 1947. pp.85. [1376.]

lxxxv. 1948. pp.86. [1487.]
lxxxvi. 1949. pp.78. [1380.]
lxxxvii. 1950. pp.89. [1581.]
lxxxviii. 1951. pp.108. [2031.]
lxxxix. 1952.
xc. 1953. Compiled by W. P. C. Tenison.
   pp.109. [2112.]
xci. 1954. pp.101. [1972.]
xcii. 1955. pp.112. [2167.]
xciii. 1956. pp.121. [2000.]
xciv. 1957. pp.107. [1750.]
xcv. 1958. pp.112. [1750.]
xcvi. 1959. pp.73. [1750.]
xcvii. 1960. pp.83. [2000.]
xcviii. 1961. pp.74. [1750.]

*in progress; earlier issues were not published
separately; the issues for 1906–1914 also form part
of vols.vi–xiv of the* International catalogue of
scientific literature, N. Zoology.

AUGUST VON PELZELN, Bericht über die leistungen in der naturgeschichte der vögel während des
jahres. Berlin.
   1879. pp.[ii].96. [1750.]
   1880. pp.78. [1500.]

BIRDS. Free public library: Special reading list:
Worcester, Mass. 1899. pp.12. [200.]

R. BOWDLER SHARPE, A hand-list of the genera and species of birds. [Nomenclator avium tum fossilium tum viventium]. [British museum (natural history):] 1899–1909. pp.xxi.303+xv.312 +xii.367+xii.391+xx.694. [25,000.]

—— General index. [By Thomas Wells]. Edited by W. R. Ogilvie-Grant. 1912. pp.v.200.

HENRY TENNYSON FOLKARD, Ornithology. A list of books about birds preserved in the reference department of the Wigan free public library. Wigan 1904. pp.15. [400.]
25 copies printed.

A LIST of bird books. Massachusetts Audubon society: Boston 1907. pp.14. [50.]

CATALOGUE of a collection of books on ornithology in the library of Frederic Gallatin, jr. New York 1908. pp.179. [1000.]
privately printed.

READING list on birds and bird study. Grosvenor library: Buffalo, N.Y. 1909. pp.15. [200.]

A LIST of books about birds in the Seattle public library. Seattle 1909. pp.11. [150.]

EVELYN THAYER and VIRGINIA KEYES, Catalogue of a collection of books on ornithology in the library of John E[liot] Thayer. Boston 1913. pp.

[ii].188. [1100.]
privately printed.

LIST of books on ornithology in the Public library of South Australia and other Adelaide libraries. Public library: Adelaide 1926. pp.40. [750.]

JOHN TODD ZIMMER, Catalogue of the Edward E. Ayer ornithological library. Field museum of national history: Publication (no.239 = Zoological series, vol.xvi): Chicago 1926. pp.[ii].x.364 +[ii].365–706. [2000.]

JAMES LEE PETERS, Check-list of birds of the world. Cambridge, Mass.

   i. [Struthio–Ieracidea]. 1931. pp.xviii.345. [3000.]

   ii. [Megapodius–Lunda]. 1934. pp.xvii.401. [3500.]

   iii. [Syrrhaptes–Geopsittacus]. 1937. pp.xiii. 311. [3000.]

   iv. [Tauraco–Hemiprocne]. 1940. pp.xii.291. [2500.]

   v. [Doryfera–Bucorvus]. 1945. pp.xi.306. [2500.]

   vi. [Galbalcyrhynchus–Campephilus]. 1948. pp.xi.259. [2000.]

REUBEN MYRON STRONG, A bibliography of birds, with special reference to anatomy, behavior, biochemistry, embryology, pathology, physiology, genetics, ecology, aviculture, economic ornithology, poultry culture, evolution, and related subjects. Field museum of natural history: Publication (nos.442, 457 = Zoological series, vol. xxv): Chicago 1939. pp.464+465–937. [20,000.]

ROBERT M[ORROW] MENGEL, A catalog of an exhibition of landmarks in the development of ornithology from the Ralph N[icholson] Ellis collection of ornithology in the university of Kansas libraries. Lawrence 1957. pp.33. [100.]

S[IDNEY] DILLON RIPLEY and LYNETTE L. SCRIBNER, Ornithological books in the Yale university library, including the library of William Robertson Coe. New Haven 1961. pp.[ix].338. [5000.]*

2. Countries

America, north

CLARENCE M[OORES] WEED, A partial bibliography of the economic relations of north american birds. New Hampshire college agricultural experiment station: Technical bulletin (no.5): Durham 1902. pp.[ii].139–179. [300.]

JOSEPH GRINNELL, A bibliography of California ornithology. Cooper ornithological club of California: Pacific coast avifauna (no.5): Santa Clara

1909. pp.166. [2000.]

— — Second instalment, to end of 1923. . . . (no.16): Berkeley 1924. pp.191. [2500.]

ALEXANDER WETMORE, A check-list of the fossil birds of north America. Smithsonian institution: Miscellaneous collections (vol.xcix, no.4 = Publications, no.3587): Washington 1940. pp.[ii].81. [750.]

Antarctic

BRIAN ROBERTS, A bibliography of antarctic ornithology. [British museum (natural history): British Graham Land expedition 1934–1937: Scientific reports (vol.i, no.9): 1941]. pp.337–367. [400.]

Australia

GREGORY M. MATHEWS, The birds of Australia. [Supplement no.4(–5)]. Bibliography of the birds of Australia. Books used in the preparation of this work. 1925. pp.[viii].149. [2000.]

H. M. WHITTELL and D. L. SERVENTY, A systematic list of the birds of Western Australia. Public library, museum and art gallery of Western Australia: Special publication (no.1): Perth 1948. pp.vi.126. [1000.]

HUBERT MASSEY WHITTELL, The literature of australian birds. Perth 1954. pp.800. [10,000.]

Ornithology

Denmark

TAGE LA COUR, Fuglebøger. Danske ornithologiske skrifter, 1648–1948. Om bøger: København 1949. pp.47. [100.]

France

RENÉ RONSIL, Bibliographie ornithologique française. Travaux publiés en langue française et en latin en France et dans les colonies françaises de 1473 à 1944. Encyclopédie ornithologique (vols.viii–ix): 1948–1949. pp.535+90. [11,607.]

Great Britain

W[ILLIAM] H[ERBERT] MULLENS and H[ARRY] KIRKE SWANN, A bibliography of british ornithology from the earliest times to the end of 1912. [1916–]1917. pp.xx.673.ff.675–691. [5000.]

W[ILLIAM] H[ERBERT] MULLENS, H[ARRY] KIRKE SWANN and F[RANCIS] C[HARLES] R[OBERT] JOURDAIN, A geographical bibliography of british ornithology. [1919–]1920. pp.viii.559. [15,000.]

[EDWARD GREY,] VISCOUNT GREY OF FALLODON, 'Our common birds'. National book league: Reader's guide: 1949. pp.20. [60.]

the bibliography proper is by William Arthur Munford.

Ornithology

RAYMOND IRWIN, British bird books. An index to british ornithology, A.D. 1481 to A.D. 1948. 1951. pp.xix.398. [7000.]

RAYMOND IRWIN, British birds and their books. Catalogue of an exhibition. National book league: [1952]. pp.38. [181.]

Scandinavia

J[OHAN] M[ARKUS] HULTH, Öfversikt af faunistiskt och biologiskt vigtigare litteratur rörande nordens fåglar. Stockholm 1899. pp.[ii].16. [850.]

Sweden

JOHAN OTTO VON FRIESEN, Öfversigt af Sveriges ornithologiska litteratur. Stockholm 1860. pp.44. [300.]

a bibliography of swedish birds.

Switzerland

THEOPHIL STUDER, Vögel. Centralkommission für schweizerische landeskunde: Bibliographie der schweizerischen landeskunde (section IV. 6. 4β): Bern 1895. pp.xiv.43. [800.]

United States

WITH the birds of Indiana. Public library commission: [Indianapolis] 1904. pp.32. [200.]

Ornithology

JOSEPH GRINNELL, A bibliography of California ornithology. Cooper ornithological club of California: Pacific coast avifauna (no.5): Santa Clara 1909. pp.166. [2000.]

— — Second installment to end of 1923 . . . (no.16): Berkeley 1924. pp.191. [2500.]

W[ALDO] L[EE] MCATEE, Index to papers relating to the food of birds by members of the Biological survey in publications of the United States Department of agriculture, 1885–1911. Department of agriculture: Bureau of biological survey: Bulletin (no.43): Washington 1913. pp.69. [200.]

EARE AMOS BROOKS, A descriptive bibliography of West Virginia ornithology. Newton Highlands, Mass. 1938. pp.iii.28. [500.]*

T[HOMAS] C[ALDERWOOD] STEPHENS, An annotated bibliography of South Dakota ornithology. Sioux City 1945. pp.[iv].28. [250.]
privately reproduced from typewriting.

— — [another edition]. Nebraska ornithologists' union: Occasional papers (no.2): Crete 1956. pp.22. [300.]

EARLE R[OSENBURG] GREENE [*and others*], Birds of Georgia. A preliminary check-list and bibliography. Georgia ornithological society: Occasional publication (no.2): Athens 1945. pp.5–111.

[1000.]

T[HOMAS] C[ALDERWOOD] STEPHENS, An annotated bibliography of Iowa ornithology. Nebraska ornithologists' union: Occasional papers (no.2): Crete 1956. pp.[iv].22. [250.]*

— — [another edition]. . . . (no.4): 1957. pp. [vi].114. [2000.]*

3. *Iconography*

CLAUS NISSEN, Schöne vogelbücher. Ein überblick der ornithologischen illustration, nebst bibliographie. Wien &c. 1936. pp.96. [554.]
100 copies printed.

JEAN ANKER, Bird books and bird art. An outline of the literary history and iconography of descriptive ornithology. University library: Copenhagen 1938. pp.xviii.251. [918.]
400 copies printed.

CLAUS NISSEN, Die illustrierten vogelbücher, ihre geschichte und bibliographie. Stuttgart 1953. pp.223. [1031.]

SACHEVERELL SITWELL, HANDASYDE BUCHANAN and JAMES FISHER, Fine bird books, 1700–1900. 1953. pp.viii.120. [500.]

DER VOGEL in buch und bild. Führer durch eine ausstellung . . . von Gesner über Naumann bis zur gegenwart. Naturhistorisches museum: Bern 1954. pp.68. [200.]

4. *Miscellaneous*

LIST of references on Audubon societies. Library of Congress: Washington 1915. ff.2. [18.]★

K. LAMBRECHT, Aves. Fossilium catalogus (I. 12): Berlin 1921. pp.[iv].104. [2000.]

THE FLIGHT of birds: 1929–1934, with select references prior to that date. Science library: Bibliographical series (no.180): 1935. ff.[i].5. [100.]★

— [second edition] . . . (no.295): 1937. ff.30. [600.]★

— — Supplement. . . . (no.788): [1963]. pp.18. [250.]★

CONTROL OF BIRD PESTS. Partially annotated bibliography. Science library: Bibliographical series (no.233): 1936. ff.11. [202.]★

— Supplement. . . . Patent literature . . . (no. 236): 1936. ff.2. [36.]★

EFFECT of hypo- and hyperthyroidism on the plumage of birds. Science library: Bibliographical series (no.272): 1936. ff.5. [73.]★

PUBLICATIONS on attracting birds. Fish and wildlife service: Wildlife leaflet (no.201): Washington 1940. pp.6. [70.]

AIDS for bird study. Fish and wildlife service: Wildlife leaflet (no.180): Washington 1941. pp.9. [170.]

ANNIE P[URDIE] GRAY, Bird hybrids. A check-list with bibliography. Commonwealth bureaux of animal breeding and genetics: Technical communication (no.13): Farnham Royal [1958]. pp.x.390. [10,000.]

ELLIS A[RDEN] HICKS, Check-list and bibliography of the occurrence of insects in birds' nests. Ames [1959]. pp.683. [17,500.]

Paleontology.

1. *Periodicals*

ROBERTO PEREZ DE ACEVEDO, Un dia . . . y otro dia mas (esquema de recopilación periodistica). Cuaderna 1. Prehistoria, paleontología. La Habana 1959. pp.20. [200.]

2. *General*

JOHANN SAMUEL SCHRÖTER, Für die litteratur und kenntniss der naturgeschichte, sonderlich der conchylien und der steine. Weimar 1782. pp. [xiv].278.[viii]+[xii].314.[viii]. [125.]

JOHANN SAMUEL SCHRÖTER, Neue litteratur und beyträge zur kenntniss der naturgeschichte, vorzüglich der conchylien und fossilien. Leipzig 1784–1787.   pp.[viii].550.[xxx] + [viii].598. [xxxiv] + [vi].614.[xxvii] + [xii].456.[xxvi]. [500.]

GOTTHELF FISCHER, Bibliographia paleonthologica animalium systematica. . . . Editio altera. Mosquae 1834. pp.[ii].viii.414. [1500.]

[ALCIDE D'ORBIGNY], Notice analytique sur les travaux zoologiques et paléontologiques de m. Alcide d'Orbigny. 1844. pp.48. [68.]

HEINRICH G[EORG] BRONN, Handbuch einer geschichte der natur. . . . Dritter band. Erster abtheilung erste[–zweite] hälfte. III. Theil. . . . Index palaeontologicus. . . . A. Nomenclator palaeontologicus, in alphabetischer ordnung. Naturgeschichte der drei reiche (vol.xv): Stuttgart 1848. pp.viii.lxxxiv.775+[iii].777–1382. [70,000.]

C[HRISTOPH] G[OTTFRIED] GIEBEL, Bericht über die leistungen im gebiete der paläontologie mit besonderer berücksichtigung der geognosie während der jahre 1848 und 1849. Berlin 1851. pp. iv.282. [1500.]

PALAEONTOLOGY. International catalogue of scientific literature (section K): Royal society.

i. 1903. pp.xiv.170. [673.]
ii. 1904. pp.viii.224. [638.]
iii. 1905. pp.viii.256. [751.]
iv. 1906. pp.viii.248. [709.]
v. 1907. pp.viii.300. [757.]
vi. 1908. pp.viii.332. [888.]
vii. 1909. pp.viii.330. [850.]
viii. 1910. pp.viii.274. [1040.]
ix. 1911. pp.viii.241. [945.]
x. 1913. pp.viii.198. [825.]
xi. 1914. pp.viii.185. [788.]
xii. 1914. pp.viii.192. [895.]
xiii. 1916. pp.viii.146. [517.]
xiv. 1919. pp.viii.130. [505.]
no more published.

FOSSILIUM catalogus. 1: Animalia. Editus a
F. Frech [parts 11–37: C. Diener; 38–47: Josef
Felix Pompeckj; 48 &c.: Werner Quenstedt].
Berlin [1934 &c.: 's Gravenhage] 1913 &c.
in progress?

K. LAMBRECHT, WERNER QUENSTEDT and ANNE-
MARIE QUENSTEDT, Palaeontologi. Catalogus bio-
bibliographicus. Fossilium catalogus (1. 72):
's-Gravenhage 1938. pp.xxii.495. [5000.]

BIBLIOGRAPHY of fossil vertebrates. Geological
society of America: [Washington] 1940 &c.

Paleontology

in progress; details of this work are entered under Vertebrates, below.

B. WOLF, Fauna fossilis cavernarum. Fossilium catalogus (I. 82, 89 &c.).

BIBLIOGRAPHY of vertebrate paleontology and related subjects. Society of vertebrate paleontology: [Chicago] 1946 &c.★
in progress; previously formed part of the society's News bulletin.

GEORGE E. FAY, A bibliography of fossil man. Southern state college: Department of sociology and anthropology: Magnolia, Ark. 1959. ff.[i].145. [2000.]★

H[EINRICH] HILTERMANN [*and others*], Bibliographie stratigraphisch wichtiger mikropaläontologischer publikationen von etwa 1830 bis 1958. Stuttgart 1961. pp.v.403. [3210.]

WERNER QUENSTEDT, Clavis bibliographica. Fossilium catalogus: Animalia (no.102): 's-Gravenhage [1963]. pp.118.

3. *Countries*
Africa

ARTHUR TINDELL HOPWOOD and JUNE PAMELA

Paleontology

HOLLYFIELD, An annotated bibliography of the fossil mammals of Africa (1742–1950). British museum (natural history): Fossil mammals of Africa (no.8): 1954. pp.[ii].194. [1000.]

America

CHARLES ROLLIN KEYES, A bibliography of north american paleontology, 1888–1892. Geological survey: Bulletin (no.121): Washington 1894. pp. vi.253. [2000.]

RAY S[MITH] BASSLER, Bibliographic index of american ordovician and silurian fossils. National museum: Bulletin (no.92): Washington 1915. pp.viii.718+iv.719–1521. [25,000.]

FRANK NICOLAS, Index to palaeontology (geological publications 1847–1916). Geological survey: Ottawa 1925. pp.x.383. [25,000.]

M[ANUEL] MALDONADO-KOERDELL, Bibliografía geológica y paleontológica de América central. Instituto panamericano de geografía e historia: Publicación (no.204): México 1958. pp.288. [2000.]

Austria

BERICHT über die österreichische literatur der zoologie, botanik und palaeontologie aus den

jahren 1850, 1851, 1852, 1853. Zoologisch-botanischer verein: Wien 1855. pp.vi.376. [1000.]
no more published.

China

PIERRE TEILHARD DE CHARDIN and PIERRE LEROY,
Chinese fossil mammals. A complete bibliography. Institut de géo-biologie (no.8): Pékin
1942. pp.iv.142. [130.]

France

[VISCOUNT É. J.] A[DOLPHE DEXMIER DE SAINT-
SIMON] D'ARCHIAC, Paléontologie de la France.
Ministère de l'instruction publique: Recueil de
rapports sur les progrès des lettres et des sciences
en France: 1868. pp.[iii].726. [1500.]
not limited to France.

Germany

C[HRISTOPH] G[OTTFRIED] GIEBEL, Deutschlands
petrefacten. Ein systematisches verzeichniss aller
in Deutschland und den angrenzenden ländern
vorkommenden petrefacten. Leipzig 1852. pp.
[ii].xiii.706. [10,000.]

C[HRISTOPH GOTTFRIED] GIEBEL, Petrefacta Ger-
maniae. . . . Repertorium zu Goldfuss' Petrefakten
Deutschlands. Ein verzeichniss aller synonymen

Paleontology

und literarischen nachweise. Leipzig 1866. pp.iv. 122. [2000.]

Great Britain

ARTHUR SMITH WOODWARD and CHARLES DAVIES SHERBORN, A catalogue of british fossil vertebrata. 1890. pp.xxxv.396. [9000.]

Italy

GIAN FRANCESCO GAMURRINI, Bibliografia dell'Italia antica. . . . Paleontologia vegetale, animale, umana. — Ricerche regionali paleontologiche e paletnologiche. Roma 1936. pp.[ii].479. [10,000.]

Physiology.

[BARON] ALBRECHT VON HALLER, Bibliotheca anatomica, qua scripta ad anatomen et physiologiam facientia a rerum initiis recensentur. Tomus I. Ad annum MDCC. [II. Ab anno MDCCI. ad MDCCLXXVI]. Tiguri 1774–1777. pp.viii.816+ [iii].870. [15,000.]
also issued with a London imprint.

CHRISTIAN LUDWIG SCHWEICKHARD, Tentamen catalogi rationalis dissertationum ad anatomiam et physiologiam spectantium ab anno MDXXXIX. ad nostra usque tempora. Tubingae 1798. pp.vi. 444. [3328.]

[CLAUDE BERNARD], Notice sur les travaux d'anatomie et de physiologie de m. Claude Bernard. [1850]. pp.38. [31.]
—— [another edition]. Notice sur les travaux de m. Claude Bernard. 1854. pp.[ii].45. [42.]

CANSTATT's jahresbericht über die leistungen in den physiologischen wissenschaften. Würzburg.
1851. pp.[ii].204. [1000.]
1852. pp.[ii].224. [1000.]
1853. pp.[ii].250. [1250.]

1854. pp.[ii].218. [1000.]
1855. pp.[ii].208. [1000.]
1856. pp.[ii].188. [750.]
1857. pp.[ii].190. [750.]
1858. pp.[ii].248. [1000.]
1859. pp.[ii].257. [1000.]
1869. pp.[ii].140. [500.]
1861. pp.[ii].260. [1000.]
1862. pp.[ii].228. [750.]
1863. pp.[ii].225. [750.]
1864. pp.[ii].222. [750.]
1865. pp.[ii].282. [1000.]

no more published; forms also the first volume of the Jahresbericht über die fortschritte der gesammten medicin &c.

[F. A. POUCHET], Notice sur les travaux de zoologie et de physiologie de m. F.-Archimède Pouchet. Rouen 1861. pp.36. [56.]
—— [another edition]. 1866. pp.50. [82.]

PAUL BERT, Revue des travaux d'anatomie et de physiologie publiés en France pendant l'année 1864. Congrès des sociétés savantes: Caen 1865. pp.61. [150.]

ADOLPH BÜCHTING, Bibliotheca anatomica et physiologica, oder verzeichniss aller . . . in den letzten 20 jahren . . . im deutschen buchhandel

erschienenen bücher und zeitschriften. Nordhausen 1868. pp.85. [850.]

[ARMAND MOREAU], Notice sur les travaux de
physiologie de m. Armand Moreau. [1868]. pp.16.
[13.]

BERICHT über die fortschritte der anatomie und
physiologie. Leipzig &c.

    1856. Herausgegeben von J. Henle und
       G[eorg] Meissner. pp.[ii].iv.650. [1250.]

    1857. pp.iv.630. [1250.]

    1858. pp.iv.648. [1250.]

    1859. pp.iv.644. [1250.]

    1860. pp.iv.624. [1250.]

    1861. Herausgegeben von J. Henle, W[il-
       helm] Keferstein und G. Meissner. pp.iv.
       468. [1000.]

    1862. pp.iv.544. [1000.]

    1863. pp.iv.452. [900.]

    1864. pp.iv.564. [1000.]

    1865. pp.iv.536. [1000.]

    1866. pp.iv.462. [900.]

    1867. pp.vi.624. [1250.]

    1868. pp.iv.512. [1000.]

    1869. Herausgegeben von J. Henle, G. Meiss-
       ner und H. Gernacher. pp.vi.538. [1000.]

    1870. pp.iv.324. [750.]

Physiology

1871. pp.v.358. [750.]
no more published.

JAHRESBERICHTE über die fortschritte der anatomie und physiologie. Leipzig 1873–1893.
details of this work are entered under Anatomy, above.

MAURICE MENDELSSOHN and CHARLES RICKET, Revue des travaux slaves de physiologie pour l'année 1885. 1886. pp.71. [100.]

JAHRESBERICHT über die fortschritte der physiologie. . . . Neue folge des physiologischen theiles der Jahresberichte von Henle und Meissner, Hofmann und Schwalbe, Hermann und Schwalbe. Bonn [vols.xi–xxi: Stuttgart; xxii: München &c.].

    i. 1892. Edited by L[udimar] Hermann. 1894. pp.viii.278. [1000.]

    ii. 1893. 1895. pp.viii.308. [1500.]

    iii. 1894. 1895. pp.vi.312. [1500.]

    iv. 1895. 1896. pp.vi.300. [1500.]

    v. 1896. 1897. pp.vi.330. [1750.]

    vi. 1897. 1898. pp.vi.306. [1750.]

    vii. 1898. 1899. pp.vi.308. [1750.]

    viii. 1899. 1900. pp.vi.324. [2000.]

    ix. 1900. 1901. pp.vi.311. [2000.]

    x. 1901. 1902. pp.vi.345. [2250.]

    — General-register zu band 1–10. 1903. pp.

[ii].140.

xi. 1902. 1903. pp.vi.341. [2250.]

xii. 1903. 1905. pp.vi.334. [2500.]

xiii. 1904. 1905. pp.viii.372. [2750.]

xiv. 1905. 1906. pp.viii.373. [2750.]

xv. 1906. 1908. pp.viii.421. [3500.]

xvi. 1907. 1909. pp.viii.496. [3500.]

xvii. 1908. Edited by L. Hermann and O. Weiss. 1910. pp.viii.562. [4000.]

xviii. 1909. 1910. pp.viii.672. [5000.]

xix. 1910. 1912. pp.viii.653. [4500.]

xx. 1911. 1913. pp.viii.556. [4000.]

— General-register zu band xi–xx. 1914. pp. [iv].231.

xxi. 1912. 1913. pp.viii.263. [2250.]

[continued as:]

Jahresbericht über die fortschritte der animali-schen physiologie.

xxii. 1913–1919. Edited by O. Weiss. 1922. pp.viii.577. [5000.]

no more published; the earlier works mentioned in the title are entered next above.

BIBLIOGRAPHIA physiologica. . . . Répertoire des travaux de physiologie de l'année. Par Ch[arles] Richet. [n.s.: Institutus bibliographicus inter-nationalis bruxellensis].

1893–1894. pp.[iv].iv.ff.184. [2750.]
1895. pp.[iv].iv.ff.110. [1500.]
1896. pp.[iv].iv.ff.52+pp.[iv].ff.72. [1750.]
new series.

i. 1897. pp.[iii].191. [2000.]
ii (no.1–2). 1897–1898. pp.64. [750.]
[*continued as:*]

Bibliographia physiologica, diario Zentralblatt für physiologie adnexa. Edidit Concilium bibliographicum sub cura Hermann Jordan [et Leo Zürcher]. Viennae &c.

third series.

i. 1905. pp.[iii].373. [4000.]
ii. 1906. pp.[ii].317. [4000.]
iii. 1907. pp.[iii].353. [4000.]
iv. 1908. pp.[iii].499. [5000.]
v. 1909. pp.[iii].318. [4000.]
vi. 1910. pp.[iii].369. [4000.]
vii. 1911. pp.[iii].592. [6000.]
viii. 1912. pp.[iii].500. [5000.]
ix. 1913. pp.[iii].606. [6000.]

no more published; only the volume and the fragment set out above were published of the second series in book form, the rest having been issued on cards.

JAHRESBERICHT über die leistungen und fortschritte in der anatomie und physiologie. Berlin 1897–1898.

Physiology

details of this work are entered under Anatomy, above.

PHYSIOLOGY [*on cover:* including experimental psychology, pharmacology, and experimental pathology]. International catalogue of scientific literature (section Q): Royal society.

    i. 1902–1903. pp.xiv.404+xii.664. [6010.]
    ii. 1904. pp.viii.619+[iii].621–1360. [9671.]
    iii. 1905. pp.viii.947. [6427.]
    iv. 1906. pp.viii.1180. [8752.]
    v. 1907. pp.viii.839+[iv].1095. [14,482.]
    vi. 1908–1909. pp.viii.832+[iv].936. [16,345.]
    vii. 1910. pp.viii.1221. [12,922.]
    viii. 1911. pp.viii.1225. [13,403.]
    ix. 1912. pp.viii.927. [10,481.]
    x. 1914. pp.viii.1155.v.182.36. [17,500.]
    xi. 1915. pp.viii.892.148.30. [12,487.]
    xii. 1917. pp.viii.890.v.111.35. [12,341.]
    xiii. 1920. pp.viii.937.v.138.30. [12,700.]
    xiv. 1921. pp.viii.830.v.132.28. [10,629.]

no more published; the 10th–14th issues contain a separate section QR: Serum physiology.

CATALOGUS van de boeken en tijdschriften geplaatst in de Laboratoria voor physiologie en histologie. Universiteit: Bibliotheek: Amsterdam 1913. pp.[ii].52. [850.]

Physiology

— 1ᵉ Vervolglijst 1919. pp.[ii].41. [700.]

SERUM physiology. International catalogue of scientific literature (section QR): Royal society.

    1914. pp.182.23[or 36]. [2625.]
    1915. pp.148.27[or 30]. [2111.]
    1917. pp.111.24[or 35]. [1600.]
    1918 [1920]. pp.138.24[or 30]. [1862.]
    1920 [1921]. pp.132.22[or 28]. [1661.]

no more published; this section was issued as part of the 10th–14th issues of both sections Q and R of the International catalogue.

PHYSIOLOGICAL abstracts. Physiological society [&c.].

    i. 1916. Edited by W. D. Halliburton. pp. ix.602. [1914.]
    ii. 1917. pp.viii.782. [3036.]
    iii. 1918. pp.vi.656. [3534.]
    iv. 1919. pp.vi.567. [3174.]
    v. 1920. pp.vi.650. [3703.]
    vi. 1921. pp.viii.729. [4396.]
    vii. 1922. pp.viii.729. [4691.]
    viii. 1923. Edited by sir W[illiam] M[addock] Bayliss. pp.[ii].592. [3923.]
    ix. 1924. Edited by J. Mellanby. pp.[ii].684. [3956.]
    x. 1925. pp.[ii].694. [4039.]

xi. 1926. pp.[ii].712. [3945.]

xii. 1927. pp.[ii].727. [4029.]

xiii. 1928. pp.[ii].746. [3830.]

xiv. 1929. pp.[ii].748. [3772.]

xv. 1930. pp.[ii].754. [3810.]

xvi. 1931. pp.[ii].801. [3914.]

xvii. [1932]. pp.[ii].847. [4022.]

xviii. [1933]. pp.[ii].826. [3847.]

xix. [1934]. pp.[iv].756. [3766.]

xx. [1935]. Edited by J. G. Priestley. pp.[iv]. 955. [5186.]

xxi. [1936]. pp.[iv].1090. [5003.]

xxii. [1937]. pp.[iv].1199. [4930.]

in January 1938 this publication was amalgamated with the biochemical section of British chemical abstracts *to form section AIII of* British chemical and physiological abstracts, *which are entered under* Chemistry, *above.*

JAHRESBERICHT über die gesamte physiologie und experimentelle pharmakologie, mit vollständiger bibliographie. Zugleich fortsetzung des Hermann–Weisschen Jahresberichts über die fortschritte der animalischen physiologie und des Maly-Spiro Andreaschschen Jahresberichts über die fortschritte der tierchemie. München &c.

i. 1920. Edited by P. Rona and K. Spiro. 1923. pp.viii.986. [9000.]

ii. 1921. 1924. pp.x.767. [7000.]

iii. 1922. 1925. pp.[iv].558+x.561–1395. [12,500.]

iv. 1923. 1925. pp.xi.900. [12,500.]

v. 1924. 1926. pp.[iv].776+xi.779–1833. [15,000.]

vi. 1925. 1929. pp.xi.1039. [15,000.]
no more published.

SILVESTRO BAGLIONI, La fisiologia. Guide bibliografiche: Roma 1923. pp.54. [252.]

H[ELEN] FLANDERS DUNBAR, Emotions and bodily changes. A survey of literature on psychosomatic interrelationships, 1910–1933. Josiah Macy, jr. foundation: New York 1935. pp.xvii.595. [2251.]

— — 1910–1945. Third edition. 1946. pp.lix. 604.

РАБОТЫ по физиологии в изданиях Академии наук . . . 1735–1934. Академия наук СССР: Москва &c. 1935. pp.65. [429.]

JOHN F[ARQUHAR] FULTON, A bibliography of two Oxford physiologists, Richard Lower . . . John Mayow. Oxford 1935. pp.62. [172.]

PHYSIOLOGIE. Revue annuelle. Dirigée par Émile F. Terroine. 1938–1939.

no more published; issued in sections, which have been entered under their subjects.

Physiology

ANNUAL review of physiology. American physiological society: Stanford university.

    i. 1939. James Murray Luck, editor. pp.vii. 705. [4000.]

    ii. 1940. pp.vii.501. [4000.]

    iii. 1941. pp.viii.784. [5000.]

    iv. 1942. pp.viii.709. [4000.]

    v. 1943. pp.viii.613. [3500.]

    vi. 1944. pp.viii.630. [3500.]

    vii. 1945. pp.viii.774. [5000.]

    viii. 1946. pp.viii.658. [3500.]

    ix. 1947. pp.vi.736. [4000.]

    x. 1948. pp.xi.552. [3000.]

    xi. 1949. pp.ix.643. [3500.]

    xii. 1950. pp.vii.609. [3500.]

    xiii. 1951. pp.xi.457. [2500.]

    xiv. 1952. pp.ix.589. [3500.]

    xv. 1953. pp.ix.558. [3000.]

    xvi. 1954. pp.ix.545. [3000.]

in progress.

FRIEDRICH HERMANN REIN [*and others*], Physiology. Office of military government for Germany: Field information agencies technical: Fiat review of german science, 1939–1946: Wiesbaden 1948.

    i–ii. Vegetative physiology. pp.[viii].224+ [iv].235. [2500.]

    iii. Animal physiology and physiology of

perception. pp.[ii].378. [5000.]
the text is in german.

советское медицинское реферативное обозрение. Нормальная и патологическая физиология, биохимия, фармакология, токсикология. Москва.

 i. Редактор Н. Т. Шутова. 1949. pp.152. [500.]

 ii. Редактор П. Д. Горизонтов. 1952. pp. 168. [500.]

in progress?

L. I. DVINYANINOV, Работы по физиологии и патологии пищеварения выполненные в лаборатории И. П. Павлова . . . с 1888 по 1934 гг. Академия наук СССР: Институт физиологии им. И. П. Павлова: Москва 1961. pp.336. [4000.]

V[LADIMIR] N[IKOLAEVICH] NIKITIN, Russian studies on age-associated physiology, biochemistry, and morphology. Public health service: Publication (no.857): Washington 1961 [1962]. pp.iv.203. [2953.]

Zoology.

Zoology

4. Nomenclature, 287.
5. Miscellaneous, 287.

1. *Periodicals*

G. C. J. VOSMAER, Welke periodica zoologica kunnen in de nederlandsche openbare bibliotheken geraadpleegd worden? Nederlandsch natuur- en geneeskundig congress: 's Gravenhage 1898. pp.vii.372.xi.

JANINA ROŻNOWSKA-FELIKSIAKOWA, Wydawnictwa ciągłe w bibliotece Instytutu zoologicznego Polskiej akademii nauk. Warszawa 1958. pp.800. [3011.]

F[REDERICK] C[HARLES] SAWYER and SYLVIA FITZGERALD, List of serial publications in the library of the Department of zoology, British museum (natural history). Second edition. 1961. pp.iv.ff.152. [1500.]*

2. *General*

1. [JEREMIAS DAVIS] REUSS, Repertorium commentationum a societatibus litterariis editarum. . . . Tom. I. Historia naturalis, generalis et zoologia. Gottingae 1801. pp.[iii].iv.74.xx.75–574. [zoology: 4000.]

Zoology

ÅRSBERÄTTELSE om nyare zoologiska arbeten och upptäckter. Kongliga vetenskaps-academien: Stockholm.

    1826. Af J. W. Dalman. pp.[ii].115.[iii]. [100.]

    1827. pp.[ii].113.[iv]. [100.]

    1828. pp.[ii].iv.138. [125.]

    1829. Af S[ven] Nilsson. pp.[ii].iv.134. [125.]

    1830. pp.vi.122. [100.]

    1831. pp.[ii].iii.108. [100.]

    1832. Af B. F. Fries. pp.viii.193. [150.]

    1833. pp.[ii].iv.123. [125.]

    1834. pp.viii.208. [200.]

    1835–1836. pp.viii.170. [300.]

    1837–1840. Af C. J. Sundewall. pp.[ii].xvi.
    582. [1500.]

    1840–1842. Af C[arl] H. Boheman. pp.

earlier issues formed part of the Academy's Årsberättelse.

ARCHIV für naturgeschichte. . . . Zweiter band [vols.lxxvi–lxxvii: II.[–vi] band; lxxviii–lxxxix: abteilung B]. Berichte über die leistungen im gebiete der natur-geschichte während des jahres [vols.vii–viii: Zoologische und botanische jahresberichte; ix–lxiii, lxxv–lxxxix: no special title; lxiv–lxxiv: Jahresberichte]. Berlin.

    ii. 1835. Herausgegeben von Ar[end] Fr[iedrich] Aug[ust] Wiegmann. pp.[iii].368.

[2000.]

iii. 1836. pp.[iii].342. [2000.]

iv. 1837. pp.[v].394. [2500.]

v. 1838. pp.vi.428. [2500.]

vi. 1839. pp.[iii].366. [2000.]

vii. 1841. Herausgegeben von W[ilhelm] F[erdinand] Erichson. pp.[iii].474. [2500.]

viii. 1842. pp.[iii].462. [2500.]

ix. 1843. pp.[iii].432. [2500.]

x. 1844. pp.[iii].443. [2500.]

xi. 1845. pp.[iii].416. [2500.]

xii. 1846. pp.[iii].443. [2500.]

xiii. 1847. pp.[iii].474. [2500.]

xiv. 1848. pp.[iii].398. [2500.]

xv. 1849. Herausgegeben von F[ranz] H[ermann] Troschel. pp.[iii].448. [2500.]

xvi. 1850. pp.[iii].468. [2500.]

xvii. 1851. pp.[iii].468. [2500.]

xviii. 1852. pp.[iii].430. [2500.]

xix. 1853. pp.[iii].412. [2500.]

xx. 1854. pp.[iii].474. [2500.]

xxi. 1855. pp.iv.486. [2500.]

xxii. 1856. pp.iv.454. [2500.]

xxiii. 1857. pp.vi.486. [2500.]

xxiv. 1858. pp.iv.535. [3000.]

xxv. 1859. pp.iv.556. [3000.]

xxvi. 1860. pp.iv.531. [3000.]

Zoology

xxvii. 1861. pp.iv.564. [3000.]

xxviii. 1862. pp.iv.571. [3000.]

xxix. 1863. pp.iv.719. [4000.]

xxx. 1864. pp.[iv].457. [2500.]

xxxi. 1865. pp.iv.682. [4000.]

xxxii. 1866. pp.iv.468. [2500.]

xxxiii. 1867. pp.iv.534. [3000.]

xxxiv. 1868. pp.iv.492. [3000.]

xxxv. 1869. pp.iv.344. [2000.]

xxxvi. 1870. pp.iv.584. [3000.]

xxxvii. 1871. pp.iv.484. [3000.]

xxxviii. 1872. pp.[iii].358. [2000.]

xxxix. 1873. pp.iv.567. [3000.]

xl. 1874. pp.iv.506. [3000.]

xli. 1875. pp.iv.517. [3000.]

xlii. 1876. pp.iv.605. [3500.]

xliii. 1877. pp.iv.574. [3500.]

xliv. 1878. pp.[iii].714. [4000.]

xlv. 1879. pp.iv.736. [4000.]

xlvi. 1880. pp.iv.820. [5000.]

xlvii. 1881. Herausgegeben von [Carl] Ed[uard] von Martens. pp.iv.646. [5000.]

xlviii. 1882. pp.[iii].690. [5000.]

xlix. [1883]. pp.iv.814. [6000.]

l. 1884. pp.iv.746. [6000.]

li. 1885. pp.[vi].272.416.214. [7500.]

lii. 1886. Herausgegeben von F[ranz] Hilgendorf. pp.vi.436.328.380. [9000.]

liii. 1887. pp.viii.450.387.406. [10,000.]

liv. 1888. pp.vi.330.228.304. [8000.]

lv. 1889. pp.vi.470.290.180. [8000.]

lvi. 1890. pp.vii.470.318.256. [9000.]

lvii. 1891. pp.vi.452.419.188. [9000.]

lviii. 1892. pp.vi.524.398.292. [10,000.]

lix. 1893. pp.vii.340.416.330. [9000.]

lx. 1894. pp.viii.546.436.338. [10,000.]

lxi. 1895. pp.x.386.912.320. [12,500.]

lxii. 1896. Herausgegeben von W. Weltner. pp.vii.428.748.290. [12,500.]

lxiii. 1897. pp.viii.376.730.506. [12,500.]

lxiv. 1898. pp.x.417.1206.229. [15,000.]

lxv. 1899. pp.ix.485.1004.526. [17,500.]

lxvi. 1900. pp.ix.368.1244.596. [17,500.]

lxvii. 1901. pp.xi.452.1472.622. [20,000.]

lxviii. 1902. pp.xi.[2690]. [25,000.]

lxix. 1903. pp.xiii.[2364]. [20,000.]

lxx. 1904. pp.xiii.[2366]. [20,000.]

lxxi. 1905. pp.xiii.[2417]. [20,000.]

lxxii. 1906. Herausgegeben von W. Weltner [part 2: Embrik Strand]. pp.xiii.[2510]. [20,000.]

lxxiii. 1907. pp.xiii.[1771]. [17,500.]

lxxiv. 1908. pp.iv.[457] + v.262.318 + vii.[724]. [17,500.]

lxxv. 1909. pp.[iii].578] + [iv.1108] + viii. [747]. [25,000.]

lxxvi. 1910. Herausgegeben von Embrik Strand. pp.[ii].252.174 + [ii].348 + [ii]. 309+486.160+[ii].258.317. [25,000.]

lxxvii. 1911. pp.[ii].258.108+[ii].360+[ii]. 325.173 + [ii].344.200 + [ii].229.344. [25,000.]

lxxviii. 1912. pp.[ii].100+[ii].164+[ii].38.+ [ii].225 + [ii].328 + [ii].444 + [ii].161 + [ii].136 + [ii].344 + [ii].285 + [ii].258 + [ii].286. [30,000.]

lxxix. 1913. pp.[ii].90+[ii].244+[ii].86+ [ii].211 + [ii].408 + [ii].384 + [ii].280 + [ii].184 + [ii].326 + [ii].250 + [ii].314 + [ii].658. [35,000.]

lxxx. 1914. pp.[ii].96 + [ii].224 + [ii].72 + [ii].109 + [ii].394 + [ii].401 + [ii].221 + [ii].202 + [ii].313 + [ii].283 + [ii].588 + [ii].448. [35,000.]

vol.i contains no bibliographical part; in vols.li–lxxv the three parts are devoted roughly to vertebrata, insecta, evertebrata respectively; in lxxvi–lxxvii these subjects are covered by parts 2, 3–5, 6; in lxxviii the twelve parts are made up thus: 1. mammalia; 2. aves; 3. reptilia, amphibia; 4. pisces; 5–9. insecta; 10. myriopoda, arachnida, prototracheata, crustacea; 11–12. ever-

tebrata &c.; in many volumes the nominal date does not correspond to the period actually covered; dates of publication are very irregular and are frequently widely separated from the periods covered; abteilung B of vols.lxxxi–lxxxix, for 1915–1923, was published only in a fragmentary form; only the following parts have been issued: lxxxi.1–4, 6–12; lxxxii.1–6, 9, 11; lxxxiii.3, 4, 7, 8, 10–12; lxxxiv.3, 5; lxxxv.1, 3, 5; lxxxvi.2, 3; lxxxvii–lxxxix.3; lxxxi–lxxxiii.3 and lxxxii–lxxxiii.4 each form a single part; various runs of several sections of this work were issued with distinct title-pages as independent series, sometimes without any indication of their origin.

LOUIS [JEAN RODOLPHE] AGASSIZ, Matériaux pour servir à une énumération aussi compète que possible des ouvrages publiés sur l'histoire naturelle . . . que je me propose d'éditer plus tard sous le titre de Bibliothèque zoologique et paléontologique. [Neuchâtel 1841–1845]. ff.[580]. [20,000.]

— — [another edition]. Nomenclator zoologicus, continens nomina systematica generum animalium tam viventium quam fossilium . . . adjectis auctoribus, libris, in quibus reperiuntur, anno editionis, etymologia et familiis, ad quas pertinent, in singulis classibus. Soloduri 1842–

1846. pp.[ii].xlii.[952]. [32,500.]

— — Nomenclatoris zoologici index universalis. 1846. pp.viii.394.

— — [supplement]. Count Friedrich August Marschall, Nomenclator zoologicus. Zoologisch-botanischer verein: Vindobonae 1873. pp.v.482. [24,000.]

REPORTS on the progress of zoology and botany, 1841, 1842. Ray society: Edinburgh [printed] 1845. pp.viii.43.348.104.xix. [zoology: 2000.]

REPORTS on zoology for 1843, 1844. Ray society: 1847. pp.viii.596. [3000.]

LOUIS [JEAN RODOLPHE] AGASSIZ, Bibliographia, zoologiæ et geologiæ. . . . Edited by H. E. Strickland [vol.iv: — and sir William Jardine]. Ray society: 1848–1854. pp.xxiii.506+[iii].492+[iii]. 658+[iii].604. [40,000.]

J. VICTOR CARUS and WILHELM ENGELMANN, Bibliotheca zoologica. Verzeichniss der schriften . . . welche in den periodischen werken enthalten und vom jahre 1846–1860 selbständig erschienen sind. Leipzig 1861. pp.x.950+xxiv.951–2144. [40,000.]

also forms a supplement to Engelmann's Bibliotheca historico-naturalis.

— — O[tto] Taschenberg, Bibliotheca zoolo-

gica. i. Verzeichniss . . . 1861–1880. [1886–]1887–
1923. pp.xx.864+viii.865–1728+iv.1729–2760
+ v.2761–3648 + vi.3649–4708 + vii.4709–5512
+viii.5513–6256+ix.6257–6620. [160,000.]

THE RECORD of zoological literature. Edited by
Albert C. L. G. [Albrecht Carl Ludwig Gotthilf
Günther.

    i. 1864. pp.vii.634. [3000.]
    ii. 1865. pp.x.798. [5000.]
    iii. 1866. pp.ix.649. [5000.]
    iv. 1867. pp.ix.678. [5000.]
    v. 1868. pp.ix.592. [5000.]
    vi. 1869. pp.x.683. [6000.]
        [continued as:]
The zoological record. [Zoological record
association].

    vii. 1870. Edited by Alfred Newton. pp.xvi.
      623. [5000.]
    viii. 1871. pp.xv.496. [5000.]
    ix. 1872. pp.xvi.495. [5000.]
    x. 1873. Edited by Edward Caldwell Rye.
      pp.xxiv.543. [5000.]
    xi. 1874. pp.xviii.557. [5000.]
    xii. 1875. pp.xviii.592. [5000.]
    xiii. 1876. pp.xviii.[568]. [5000.]
    in vol.xiii and all subsequent volumes each sec-
    tion of the volumes has a distinct pagination.

xiv. 1877. pp.xxiii.[615]. [5000.]

xv. 1878. pp.xxvi.[668]. [5000.]

xvi. 1879. pp.xxiv.[698]. [5000.]

xvii. 1880. pp.xxiv.[682]. [5000.]

xviii. 1881. pp.xxvi.[713]. [6000.]

xix. 1882. pp.xxviii.[728]. [6000.]

xx. 1883. pp.xxxii.[695]. [6000.]

xxi. 1884. Edited by F. Jeffrey Bell. pp.xxxii. [720]. [6000.]

xxii. 1885. pp.xxxvi.[714]. [6000.]

xxiii. 1886. Edited by Frank E. Beddard. [Zoological society]. pp.xliii.[829]. [7500.]

xxiv. 1887. pp.xl.[868]. [7500.]

xxv. 1888. pp.xlii.[865]. [7500.]

xxvi. 1889. pp.xliv.[846]. [7500.]

xxvii. 1890. pp.xlii.[794]. [7500.]

xxviii. 1891. Edited by D. Sharp. pp.xliv. [938]. [7500.]

xxix. 1892. pp.xlii.[974]. [7500.]

xxx. 1893. pp.l.[1029]. [7500.]

xxxi. 1894. Zoological society. pp.lii.[912]. [7500.]

xxxii. 1895. pp.liv.[1096]. [7500.]

xxxiii. 1896. pp.l.[837]. [7500.]

xxxiv. 1897. pp.lii.[1055]. [7500.]

xxxv. 1898. pp.liv.[939]. [7500.]

xxxvi. 1899. pp.lv.[1046]. [7500.]

xxxvii. 1900. pp.lviii.[1120]. [7500.]

xxxviii. 1901. pp.lx.[1032]. [7500.]

xxxix. 1902. pp.lxii.[1195]. [10,000.]

xl. 1903. pp.lxiv.[1203]. [10,000.]

xli. 1904. pp.lxv.[1202]. [10,000.]

xlii. 1905. pp.lxv.[1201]. [10,000.]

   *vols. xliii–li form also the 6th–14th annual
   issues of section N of the International
   catalogue of scientific literature:*

xliii. 1906. pp.xii.56.[1328]. [10,000.]

xliv. 1907. pp.xii.56.[1465]. [12,500.]

xlv. 1908. pp.xii.56.[1395]. [12,500.]

xlvi. 1909. pp.xii.56.[1350]. [12,500.]

xlvii. 1910. pp.xii.56.[1349]. [12,500.]

xlviii. 1911. pp.xii.56.[1242]. [12,500.]

xlix. 1912. pp.xii.56.[1318]. [12,500.]

l. 1913. pp.xii.56.[1295]. [12,500.]

li. 1914. pp.viii.iii–vi.56.[867]. [7500.]

   *this volume was first issued provisionally:
   pp.[iv].vi. &c.*

lii. 1915. pp.vi.[746]. [7500.]

liii. 1916. pp.iv.[690]. [5000.]

liv. 1917. pp.iv.[591]. [5000.]

lv. 1918. pp.iv.[624]. [5000.]

lvi. 1919. pp.iv.[645]. [5000.]

lvii. 1920. pp.iv.[708]. [7500.]

lviii. 1921. Superintended by W. L. Sclater.

Zoology

pp.viii.[747]. [7500.]

lix. 1922. pp.vi.[988]. [10,000.]

lx. 1923. pp.vi.[867]. [7500.]

lxi. 1924. pp.vi.[883]. [7500.]

lxii. 1925. pp.vii.[1059]. [10,000.]

lxiii. 1926. pp.vii.[1165]. [10,000.]

lxiv. 1927. pp.vii.[1208]. [12,500.]

lxv. 1928. pp.vii.[1248]. [12,500.]

lxvi. 1929. pp.vii.[1199]. [12,500.]

lxvii. 1930. pp.vii.[1297]. [12,500.]

lxviii. 1931. pp.vii.[1284]. [12,500.]

lxix. 1932. pp.vii.[1275]. [12,500.]

lxx. 1933. pp.vii.[1301]. [12,500.]

lxxi. 1934. pp.vii.[1459]. [15,000.]

lxxii. 1935. pp.vii.[1452]. [13,500.]

lxxiii. 1936. pp.vii.[1491]. [14,000.]

lxxiv. 1937. pp.vii.[1505]. [14,000.]

lxxv. 1938. Edited by Malcolm A. Smith.
pp.vii.[1383]. [13,000.]

lxxvi. 1939. pp.v.[1410]. [13,501.]

lxxvii. 1940. pp.v.[900]. [7992.]

lxxviii. 1941. pp.v.[881]. [8399.]

lxxix. 1942. pp.v.[846]. [7767.]

lxxx. 1943. pp.v.[854]. [7405.]

lxxxi. 1944. pp.v.[882]. [7360.]

lxxxii. 1945. pp.v.[1292]. [11,795.]

lxxxiii. 1946. pp.v.[1261]. [12,884.]

lxxxiv. 1947. pp.v.[1356].

lxxxv. 1948. pp.v.[1262].

lxxxvi. 1949. pp.v.[1408].

lxxxvii. 1950. Edited by G. B. Stratton.
pp.v.[1506].

lxxxviii. 1951. pp.v.[1592].

lxxxix. 1952. pp.v.[1446].

xc. 1953. pp.[ii].vi.[1673]. [15,000.]

xci. 1954. pp.[ii].vi.[1640]. [15,000.]

xcii. 1955. pp.[ii].vi.[1751]. [16,000.]

xciii. 1956. pp.[ii].vi.[2167]. [16,000.]

xciv. 1957. pp.[ii].vi.[1952]. [15,000.]

xcv. 1958. pp.[ii].vi.[2244]. [17,000.]

xcvi. 1959. 1962. pp.[ii].vi.[1326]. [18,000.]

xcvii. 1960. 1963. pp.[ii].vi.[1587]. [20,000.]

in progress.

[HENRI] MILNE EDWARDS, Rapport sur les progrès récents des sciences zoologiques en France. Ministère de l'instruction publique: Recueil de rapports sur les progrès des lettres et des sciences en France: 1867. pp.[iii].499. [1000.]

CATALOGUE of the library of the Zoological society. 1872. pp.[ii].155. [2250.]

— Fifth edition. 1902. pp.[iv].856. [12,500.]

Zoology

ZOOLOGISCHER jahresbericht. Zoologische station zu Neapel: Leipzig [1884–1913: Berlin].

1879. Redigirt von J. Victor Carus. pp.xii. 1250. [3000.]

1880. pp.ix.384+iv.436+[v].116+iv.294. [5000.]

1881. Redigirt von J. V. Carus and P. Mayer. pp.x.336 + vii.420 + [v].142 + vi.314. [6000.]

1882. Redigirt von J. V. Carus, Paul Mayer und Wilh. Giesbrecht. pp.viii.334+[iii]. 478+[iii].148+iv.304. [6000.]

1883. Redigirt von P. Mayer und W. Giesbrecht. pp.viii.292+[iii].586+[iii].114+ iv.334. [6000.]

1884. pp.viii.358 + [iii].586 + [iii].142 + iv. 414. [6000.]

1885. pp.viii.258.93 + [iii].620 + [iii].140 + iv.336. [6000.]

1886. Redigirt von P. Mayer. pp.vi.[474]. [1250.]

1887. pp.iv.[475]. [1250.]

1888. pp.iv.[555]. [1500.]

1889. pp.iv.[540]. [1500.]

1890. pp.iv.[513]. [1500.]

1891. pp.v.[587]. [1500.]

1892. pp.v.[620]. [1500.]

1893. pp.v.[578]. [1500.]

1894. pp.v.[554]. [1500.]

1895. pp.v.[574]. [1500.]

1896. pp.v.[531]. [1500.]

1897. pp.v.[530]. [1500.]

1898. pp.vi.[500]. [1500.]

1899. pp.vi.[488]. [1750.]

1900. pp.viii.[475]. [1750.]

1901. pp.viii.[499]. [1750.]

1902. pp.viii.[584]. [2000.]

1903. pp.viii.[572]. [2000.]

1904. pp.viii.[623]. [2000.]

1905. pp.viii.[660]. [2000.]

1906. pp.viii.[606]. [2000.]

1907. pp.viii.[598]. [2000.]

1908. pp.viii.[577]. [2000.]

1909. pp.viii.[672]. [2000.]

1910. pp.viii.[641]. [2250.]

1911. pp.viii.[615]. [2250.]

1912. Redigirt von P. Mayer und J. Gross. pp.viii.[604]. [2250.]

1913. Redigiert von J. Gross. 1924. pp.x.588. [3000.]

no more published.

G[ILLIS JANSE], Catalogus der bibliotheek van het Koninklijk zoölogisch genootschap natura artis magistra te Amsterdam. 1881. pp.xii.563.

Zoology

[4381.]

[J. E. HARTING], Catalogue of the books in the Department of zoology. Third edition. British museum (natural history): 1884. pp.[iv].399. [5500.]

C. EKAMA, Catalogue de la bibliothèque. . . . Troisième livraison: zoologie. Fondation Teyler: Harlem 1886. pp.[iii].179–309. [750.]

BIBLIOTHEK des professors . . . Ludwig von Graff in Graz. 1891. pp.xv.337. [12,235.]

ZOOLOGISCHER anzeiger. [Litteratur]. Leipzig.
    xiv. 1891. Herausgegeben von J. Victor
        Carus. pp.344. [6000.]
    xv. 1892. pp.336. [6000.]
    xvi. 1893. pp.568. [10,000.]
    xvii. 1894. pp.558. [10,000.]
    xviii. 1895. pp.540. [10,000.]
        [continued as:]

Bibliographia zoologica (diario 'Zoologischer anzeiger' adnexa). Lipsiae [vols.xxiv: Zürich].
    i. Edidit J. Victor Carus. 1896. pp.iv.680.
        [8773.]
    ii. 1897. pp.ix.xvi.659. [9702.]
    iii. 1898. pp.iv.575. [8389.]
    iv. 1899. pp.iv.612. [8138.]

v. 1900. pp.[iii].588. [8151.]

vi. 1901. pp.[iii].536. [7421.]

vii. 1902. pp.iv.592. [8032.]

viii. 1903. Ediderunt J. V. Carus et H. H. Field. pp.[iii].560. [8863.]

ix. 1904. Edidit Herbert Haviland Field. pp.[iii].512. [8052.]

x. 1904. pp.[iv].440. [5272.]

xi. 1905. pp.[iii].472. [6114.]

xii. 1906. pp.[iii].384. [5238.]

xiii. 1907. pp.[iii].480. [6102.]

xiv. 1908. pp.[iii].480. [6188.]

xv. 1909. pp.[iii].480. [7368.]

xvi. 1909. pp.[iii].480. [6934.]

xvii. 1910. pp.[iii].468. [6870.]

xviii. 1910. pp.[iii].480. [7199.]

xix. 1911. pp.[iii].480. [7520.]

xx. 1911. pp.[iii].480. [7173.]

xxi. 1912. pp.[iii].480. [7248.]

xxii. 1912. pp.[iii].480. [7112.]

xxiii. 1912. pp.[iii].480. [6992.]

xxiv. 1913. pp.[iv].478. [7371.]

xxv. 1914. pp.[iii].480. [6908.]

no more published; issues before 1891 appeared as part of the Zoologischer anzeiger.

CATALOG der handbibliothek des Königlichen zoologischen und anthropologisch-ethnographischen museums in Dresden. Berlin 1898. pp.xxv. 287. [zoology: 4000.]

CHILDREN's reading-list animals. Bibliography (no.6): Boston 1899. pp.24. [500.]

ZOOLOGY. International catalogue of scientific literature (section N): Royal society.

    i. 1904. pp.xvi.368+[iii].369–1528. [5918.]

    ii. 1905. pp.viii.424+[ii].xi.1009+xi.1011–1526. [7131.]

    iii. 1905–1906. pp.viii.432+xi.963+xi.438. [7313.]

    iv. 1906. pp.viii.405+xi.851+xi.384. [6473.]

    v. 1907. pp.xi.1259+viii.548. [7000.]

the remaining issues, the 6th–14th, were combined with vols.xliii–li of The zoological record *as above.*

H. W. RUTHERFORD, Catalogue of the library of Charles Darwin now in the Botany school, Cambridge. Cambridge 1908. pp.xiv.91. [1750.]

MARGARET DONAN, Animals of our zoo. A selected list of books about wild animals. Public library: St. Louis 1919. pp.7. [60.]

CHARLES DAVIES SHERBORN, Index animalium sive index nominum quae ab A.D. MDCCLVIII generibus et speciebus animalium imposita sunt. Cantabrigiae.

Zoology

Sectio prima. 1758–1800. 1902. pp.lix.1195. [60,000.]

Sectio secunda. 1801–1850.

    [A–B].   1922[–1924].   pp.cxxxix.944. [50,000.]

    C. [1924–1925]. pp.[iii].945–1771. [45,000.]

    D–G. [1925–1926].   pp.[iii].1773–2880. [50,000.]

    H–L. [1927]. pp.2881–3746. [47,500.]

    M–N. [1928]. pp.3747–4450. [40,000.]

    O–P. [1929]. pp.4451–5349. [50,000.]

    Q–R. [1930–1931]. pp.5349–6358. [55,000.]

    T–Z. [1931–1932]. pp.6359–7056. [42,500.]

    Epilogue, additions . . . and corrections, and index to trivialia. 1932[–1933]. pp.vii.cxxxiii–cxlviii.1098. [6000.]

[SPENCER SAVAGE], Catalogue of the printed books and pamphlets in the library of the Linnean society. . . . New edition. 1925. pp.[v].860. [17,500.]

J. ARTHUR THOMSON, What to read on zoology. Public libraries: Leeds 1928. pp.39. [250.]

F[RANCIS] P[ITCHER] ALLEN, A check list of periodical literature and publications of learned societies of interest to zoologists in the university of Michigan libraries. University of Michigan:

Zoology

Museum of zoology: Circular (no.2): Ann Arbor 1935. pp.3–83. [1300.]

this is a new edition of Robert W. Hegner, 'Bibliography of periodical literature and publications [&c.]', Eighteenth annual report [of the] Michigan academy of science *(1916), pp.106–118.*

READERS' guide to books on biology, botany and zoology. Library association: County libraries section [no.17]: 1938. pp.[iv].27. [600.]

SHEFFIELD AIREY NEAVE, Nomenclator zoologicus. A list of the names of genera and subgenera in zoology from the tenth edition of Linnaeus 1758 to the end of 1935. Zoological society: 1939–1940. pp.xiv.957+[iv].1025+[iii].1065+[iii].758. [225,000.]

ROGER C[LETUS] SMITH, Guide to the literature of the zoological sciences. Minneapolis 1942. ff.[i].vii.pp.128. [250.]*
— — Sixth edition. 1962. pp.[ii].xii.232. [1000.]*

ERIKA BOTHE, Tiere sehen uns an. Ein Bücherverzeichnis. Zentralinstitut für bibliothekswesen: Leipzig [1955]. pp.56. [300.]

[FRANZ XAVER PRÖLL], Von der tierschau zum tiergarten. Stadtbibliothek: Ausstellungs-katalog

Zoology

3. Countries

America

BIOLOGICAL survey. Bulletin, circulars and north american fauna of this bureau. Superintendent of documents: Price list (no.39): Washington [1909]. pp.8. [100.]

frequently re-edited under various titles.

[HOWARD H. PECKHAM, HELEN T. GAIGE and CARL L. HUBBS], Ichthyologia et herpetologia americana. A guide to an exhibition in the William L. Clements library illustrating the development of knowledge of american fishes, amphibians and reptiles. University of Michigan: William L. Clements library: Bulletin (no.xxv): Ann Arbor 1936. pp.[ii].22. [46.]

Asia

BERNARD J. CLIFTON, Zoology. Subject index to bibliographies on southwestern Asia: Subject index: Coral Gables, Fla. 1960. pp.vii.83. [6000.]

Austria

BERICHT über die österreichische literatur der zoologie, botanik und palaeontologie aus den

Zoology

jahren 1850, 1851, 1852, 1853. Zoologisch-botanischer verein: Wien 1855. pp.vi.376. [1000.]
no more published.

Brazil

MERCEDES MAIA, Bibliografia zoológica brasileira. Boletim bibliográfico: [Rio de Janeiro] 1953 &c.

BIBLIOGRAFIA brasileira de zoologia. Instituto brasileiro de bibliografia e documentação: Rio de Janeiro.

    1950–1955. pp.346.125. [2969.]
    1956–1958. pp.192. [1226.]
in progress?

Denmark

C[HRISTIAN] C[ARL] A[UGUST] GOSCH, Udsigt over Danmarks zoologiske literatur. Kjøbenhavn 1870–1878. pp.xiii.346+[iii].602+xiii.602+viii.558. [3000.]
the first volume was originally published under the title of Populær fremstilling af de videnskabelige grundsætninger for . . . zoologiens studium, *but a cancel titlepage was afterwards issued bringing it into line with the remaining volumes.*

SVEND DAHL, Bibliotheca zoologica danica, 1876–1906. København 1910. pp.xxiii.262. [1629.]

Zoology

Estonia

ZOOLOOGOLINE kirjandus 1945–1959. Eesti NSV teaduste akadeemia: Zooloogia ja botaanika instituut: Tartu 1961. pp.104. [750.]

Finland

BIBLIOTHECA zoologica Fenniae. Societas pro fauna et flora fennica: Acta (vol.xxiv, no.1): Helsingfors 1909. pp.xii.361. [6000.]

Germany

WILHELM BLASIUS, Die faunistische litteratur Braunschweigs und der nachbargebiete mit einschluss des ganzen Harzes. Braunschweig 1891. pp.239. [2250.]

JOHANNES FICKEL, Die litteratur über die tierwelt des königreichs Sachsen. Dresden 1893. pp.44. [664.]

Great Britain

JOHN SMART, Bibliography of key works for the identification of the british fauna and flora. Association for the study of systematics in relation to general biology: Publication (no.1): 1942. pp.viii.105. [750.]

— — Second edition. Edited by John Smart and George Taylor. 1953. pp.xi.126. [1250.]

Zoology

Hungary

A MAGYAR állattani irodalom ismertetése. [Kir.] Természettudományi társulat: Budapest.
- [i]. 1870–1880. Eredeti források alapján irta Daday Jano. 1882. pp.vii.coll.190. [750.]
- [ii]. 1881–1890. 1891. pp.viii.coll.256.pp. 257–264.coll.265–308. [1902.]
- iii. 1891–1900. Szerkesztette Szilády Zoltán. 1903. pp.viii.coll.512. [3932.]

Indochina

A. PÉTELOT, Analyse des travaux de zoologie et de botanique concernant l'Indochine publiés en 1929. Direction générale de l'instruction publique: Hanoi 1930. pp.22. [100.]

Italy

BIBLIOTECA della zoologia e anatomia comparata in Italia. Rivista bibliografica bimensile per Lorenzo Camerano e Mario Lessona. Torino.
- i. 1878. pp.111. [150.]
- ii. 1879. pp.[ii].96. [150.]
- iii. 1880. pp.[ii].37. [135.]

New Guinea

SELECTED list of references to publications in

Zoology

the Science library on the fauna and flora of New Guinea and the Melanesian Islands. Science library: Bibliographical series (no.705): 1951. ff.2. [23.]*

Philippines

CASTO DE ELERA, Catálogo sistemático de toda la fauna de Filipinas. Manila 1895–1896. pp.ix.701 +[iii].676+[iv].942.lxiv. [75,000.]

Poland

ANTONI JAKUBSKI and MARJI DYRDOWSKIEJ, Bibljografja fauny polskiej do roku 1880. Polska akademja umiejętnosci: Prace monograficzne Komisji fizjograficznej (vol.iii–iv): Krakowie 1927–1928. pp.470+[iii].384. [12,309.]

Russia

FRIEDRICH THEODOR KÖPPEN [FEDOR PETROVICH KEPPEN], Bibliotheca zoologica russica. Litteratur über die thierwelt Gesammtrusslands bis zum jahre 1885 incl. . . . Band 1. Allgemeiner theil. Императорская академія наукъ: С.-Петербургъ 1905–1908. pp.xvi.552+[ii].vi.366+[iii]. 367–593. [15,000.]
no more published.

Zoology

Scotland

ALEXANDER HUNTER, A selected list of books on wild animals of the highlands. Corporation public libraries: Glasgow [1951]. pp.23. [200.]

Silesia

BIBLIOGRAPHIE der schlesischen zoologie. Breslau.

[i]. Von Ferdinand Pax und H[ildegard] Tischbierek. Historische kommission für Schlesien: Schlesische bibliographie (vol. v): 1930. pp.xii.520. [6852.]

ii. 1928–1935. Von F. Pax. pp.x.178. [1929.]

iii. 1935–1950. Bibliografia zoologii Śląska. Państwowe wydawnictwo naukowe: Wrocław 1957. pp.xvi.184. [1953.]

Switzerland

A[TTILIO] LENTICCHIA, Bibliografia sulla fauna della Svizzera italiana. Centralkommission für schweizerische landeskunde: Bibliographie der schweizerischen landeskunde (section IV.6.I.c): Bern 1894. pp.iv.8. [150.]

F. ZSCHOKKE, Seenfauna. Centralkommission für schweizerische landeskunde: Bibliographie der schweizerischen landeskunde (section IV.6.2):

Zoology

Bern 1897. pp.[x].24. [250.]

TH. STUDER [*and others*], Crustacea, bryozoa, annelida, rotifera, turbellaria, spongien und hydroiden, protozoa. Centralkommission für schweizerische landeskunde: Bibliographie der schweizerischen landeskunde (section IV.6.9): Bern 1898. pp.viii.27. [500.]

Turkey

WOLFGANG NEU and HANS KUMMER LÖWE, Bibliographie der zoologischen arbeiten über die Türkei und ihre grenzgebiete. Leipzig 1939. pp.xii.62. [1100.]

United States

HERBERT OSBORN, Bibliography of Ohio zoology. Ohio biological survey: Bulletin (no.23 = vol.iv, no.8): Columbus 1930. pp.[ii].353–410. [1500.]

F[REDERICK] G[ORDON] RENNER *and others*], A selected bibliography on management of western ranges, livestock, and wildlife. Department of agriculture: Miscellaneous publication (no.281): Washington 1938. pp.ii.468. [8274.]

Zoology

4. Nomenclature

FRANCIS HEMMING and DIANA NOAKES, *edd.* Official list of works approved as available for zoological nomenclature. International trust for zoological nomenclature.

    i. 1958. pp.xi.12. [38.]
in progress.

FRANCIS HEMMING and DIANA NOAKES, *edd.* Official index of rejected and invalid works in zoological nomenclature. International trust for zoological nomenclature.

    i. 1958. pp.xi.14. [58.]
in progress.

5. Miscellaneous

[F. É. GUÉRIN-MÉNEVILLE], Liste des principaux travaux zoologiques de m. F.-Éd. Guérin-Méneville. [*c.*1840]. pp.8. [30.]
  — — [another edition]. Notice sur les travaux de zoologie. [1847]. pp.16.

[ALCIDE D'ORBIGNY], Notice analytique sur les travaux zoologiques et paléontologiques de m. Alcide d'Orbigny. 1844. pp.48. [68.]

[ÉMILE BLANCHARD], Notice sur les travaux d'anatomie et de zoologie de m. Émile Blanchard. 1850. pp.[ii].36. [54.]

[C. L. J. L. BONAPARTE], Notice sur les travaux zoologiques de m. Charles-Lucien Bonaparte. 1850. pp.35. [87.]

[ALCIDE D'ORBIGNY], Notice analytique sur les travaux de zoologie de m. Alcide d'Orbigny. Corbeil 1850. pp.[ii].47. [45.]

[J. L. A. DE QUATREFAGES DE BRÉAU], Notice sur les travaux zoologiques et anatomiques de m. A. de Quatrefages. [1850]. pp.56. [78.]
—— [another edition]. Notice sur les travaux scientifiques [&c]. 1883. pp.12. [17.]

[F. A. POUCHET], Notice sur les travaux de zoologie et de physiologie de m. F.-Archimède Pouchet. Rouen 1861. pp.36. [56.]
—— [another edition]. 1866. pp.50. [82.]

[PAUL FISCHER], Notice sur les travaux zoologiques de m. P. Fischer. 1868. pp.16. [100.]
—— [another edition]. 1875. pp.47. [200.]
——— Supplément. 1876. pp.6. [8.]

GUELFO CAVANNA, Elementi per una bibliografia italiana intorno all'idrofauna agli allevamenti degli animali acquatici e alla pesca. Firenze 1880. pp.viii.170. [2000.]

CH[ARLES] WARDELL STILES and ALBERT HASSALL, Index-catalogue of medical and veterinary zoology. Authors. Department of agriculture: Bureau of animal industry: Bulletin (no.39): Washington 1902–1912. pp.2766. [60,000.]

— — [new edition]. By Albert Hassall and Margie Potter. 1932–1952. pp.5711. [200,000.]

— — — Supplement.

1. A–B. By Mildred A. Doss . . . and Judith M. Humphrey. 1953. pp.[ii].317. [10,000.]
2. A–C. 1954. pp.317–457. [4000.]
3. A–I. 1955. pp.[ii].459–844. [12,000.]
4. A–K. By M. A. Doss . . . J. M. Humphrey . . . and Dorothy B. Segal. 1955. pp.[ii]. 845–970. [4000.]
5. A–Q. 1956. pp.[ii].971–1373. [12,000.]
6. A–Z. 1956. pp.[ii].1375–1787. [13,000.]
7. A–Z. 1957. pp.[ii].1789–2139. [7500.]
8. A–Z. 1958. pp.[ii].337. [7500.]
9. A–Z. 1959. pp.[ii].322. [7000.]
10. A–Z. 1960. pp.[ii].357. [7500.]
11. A–Z. 1961. pp.[ii].409. [8500.]
12. A–Z. 1962. pp.[ii].353. [7500.]
13. A–Z. 1963. pp.[ii].314. [6500.]

— — Subjects. Hygienic laboratory: Bulletin (no.37 &c.): 1908 &c.

in progress.

PAUL WEISS, Entwicklungsphysiologie der tiere. Wissenschaftliche forschungsberichte: Naturwis-

senschaftliche reihe (vol.xxii): Dresden &c. 1930.
pp.xi.138. [500.]

THEODORE C[EDRIC] RUCH, Bibliographia pri-
matologica. A classified bibliography of primates
other than man. Yale medical library: Historical
library: Publication (no.4 &c.): Springfield, Ill.
　　i. Anatomy, embryology & quantitative
　　　morphology; physiology, pharmacology
　　　& psychobiology; primate phylogeny &
　　　miscellanea.... (no.4): 1941. pp.xxvii.243.
　　　[4630.]
no more published.

SOME references to the care, breeding and
maintenance of laboratory animals. Science
library: Bibliographical series (no.674): [1949].
ff.3. [35.]*
　— Further references [&c.] ... (no.691): 1950.
ff.3. [40.]*

J. A. MCCORMICK, Radioisotopes in animal physiology. A literature search. Atomic energy commission: Technical information service: Oak
Ridge, Tenn. 1958. pp.118. [983.]*

Animals

Acalephae.

L[OUIS JEAN RODOLPHE] AGASSIZ, Nomina systematica generum acalepharum. [Soloduri 1842–1846]. pp.iv.7.4. [350.]

Ammoneae.

F. FRECH, Ammoneae deronicae (clymeniidae, aphyllitidae, gephyroceratidae, cheiloceratidae). Fossilium catalogus (1.1): Berlin 1913. pp.42. [600.]

Ammonoidea.

C. DIENER, Ammonoidea permiana [neo–cretacea]. Fossilium catalogus (1. 14, 29): Berlin 1921, 1925. pp.36+[ii].244. [6000.]

Amphibia.

THE ZOOLOGICAL record [IV; *afterwards:* XVI; XIV; XV]. Reptilia [*afterwards:* and batrachia; Amphibia and reptilia; Amphibia]. 1876 &c.
    xc. 1953. Compiled by Alice G. C. Grandison and W. E. Swinton. pp.70. [600.]
    xci. 1954. pp.74. [600.]
    xcii. 1955. pp.79. [500.]

xciii. 1956. pp.106. [700.]

xciv. 1957. Compiled by A. G. Grandison and Marcia E. Edwards. pp.100. [800.]

xcv. 1958.

xcvi. 1959. Compiled by A. G. C. Grandison, M. A. Edwards and Pauline Armitage [Curds]. pp.59. [900.]

xcvii. 1960. pp.130. [2000.]

in progress; details of the earlier volumes are entered under Reptiles, below.

H. FISCHER–SIGWART, Reptilien und amphibien. Centralkommission für schweizerische landeskunde: Bibliographie der schweizerischen landeskunde (section IV. 6. 5γ): Bern 1898. pp.xii.27. [500.]

[OSKAR KUHN], Amphibia. Fossilium catalogus (1, 61, 84): Berlin 1933, 1938. pp.[ii].60.114.19.26. [4000.]

HOBART M. SMITH and EDWARD H[ARRISON] TAYLOR, An annotated checklist and key to the amphibia of Mexico. Smithonian institution: United States national museum: Bulletin (no.194): Washington 1948. pp.iv.118. [600.]

Amphineura.

G. HABER, Gastropoda, amphineura et scaphopoda jurassica. Fossilium catalogus (1. 53, 65 &c.): Berlin 1932 &c.

in progress?

Anatidae.

MARQUIS NAGAMICHI KURODA, A bibliography of the duck tribe, anatidæ, mostly from 1926 to 1940, exclusive of that of dr. Phillips's work. Tokyo 1942. pp.[viii].853. [6539.]

Anguilla.

[A. G. HORNYOLD], Vingt années de recherches sur l'anguille . . . par Alfonso Gandolfi Hornyold. Lugano [printed] 1936. pp.23. [223.]

REFERENCES to the electric organ of the electric eel, electrophorus electricus Linnaeus = gymnotus electricus, 1931–1945. Science library: Bibliographical series (no.675): 1949. single leaf. [14.]*

Annelids.

OLGA HARTMAN, Literature of the polychaetus annelids. . . . Vol. 1. Bibliography. Los Angeles 1951. pp.vi.290. [4000.]*

Anthozoa.

J. FELIX, Anthozoa palaeocretacea [cenomanica,

neocretacea, eocaenica et oligocaenica, miocaenica, pliocaenica et plistocaenica]. Fossilium catalogus (1. 5–7, 28, 35, 44): Berlin 1914–1929. pp.273+296+[ii].297–668. [17,500.]

Antiarchi.

w. GROSS, Antiarchi. Fossilium catalogus (1. 57): Berlin 1932. pp.40. [400.]

Apes.

HERMANN VOSS, Bibliographie der menschenaffen (schimpanse, orang, gorilla). Jena 1955. pp. viii.163. [2000.]

Aphelinus mali.

MERILL ARTHUR YOTHERS, An annotated bibliography on *Aphelinus mali* (Hald.), a parasite of the woolly apple aphid, 1851–1950. Department of agriculture: Bureau of entomology and plant quarantine (E-861): Washington 1953. pp.61. [310.]

Aphis.

CH[ARLES] DES MOULINS, Note bibliographique sur les pucerons. Bordeaux [1869]. pp.4. [15.]

Apiculture.

AUGUSTO DE KELLER, Elenchus librorum de

apium cultura. Milano 1881. pp.[ii].iii.224. [2250.]
*a copy in the John Crerar library, Chicago, contains
additions in ms.*

ADOLPH BÜCHTING, Bibliographie für bienen-
freunde, oder verzeichniss der in bezug auf die
bienen von 1700 bis mitte 1861 in Deutschland
und der Schweiz erschienenen bücher und zeitschriften. Nordhausen 1861. pp.75. [550.]

BEE culture investigations. Publications of the
United States Department of agriculture. Super-
intendent of documents: Price list (no.66):
Washington 1915. pp.4. [26.]

CATALOGUE of books on gardening, poultry &
bees. Municipal libraries: Leicester 1927. pp.15.
[Bees: 19.]
— [another edition]. 1933. pp.15. [18.]

H. J. O. WALKER, Descriptive catalogue of a
library of bee-books. Budleigh Salterton 1929.
pp.144. [1500.]

VAJEN E[ILLEEN] HITZ and INA L[OUWE] HAWES,
List of publications on apiculture contained in the
U.S. Department of agriculture library and in
part those contained in the Library of Congress.
Department of agriculture: Bibliographical contributions (no.21): Washington 1930. pp.[iii].218.
[3000.]

A LIST of the publications on apiculture contained in the dr. Charles C. Miller agricultural library. University of Wisconsin: College of agriculture: Madison, Wis. 1936. ff.iii.pp.283. [4250.]*

[E. S. CHERNUIKOVA], Что читать по пчеловодству. Краткий аннотированный список литературы. Государственная библиотека СССР им. В. И. Ленина: Москва 1943. pp.8. [20.]

ARTIFICIAL insemination of queen bees. Science library: Bibliographical series (no.639): 1947. ff.2. [20.]*

[CATHERINE L. DICKSON], Catalogue of the [John W.] Moir library. Scottish beekeepers' association. Stirling [printed] 1950. pp.193. [1500.]

LARS FRYKHOLM, Förteckning över bitidskrifter. Supplement. [Uppsala 1952]. pp.[iv]. [100.]
the main work was published in Bitidningen *(1950).*

APICULTURAL abstracts. Bee research association.
    i–xii. 1950–1961.
    xiii. 1962. pp.183. [800.]
    xiv. 1963. pp.192. [900.]
in progress.

BEES and beekeeping. Surrey county library:
Book list (no.19): [Esher 1960]. pp.[iv].12. [75.]*

Arachnida.

L[OUIS JEAN RODOLPHE] AGASSIZ, Nomina systematica generum arachnidarum. [Soloduri 1842–
1846]. pp.iv.14. [500.]

T[AMERLAN] THORELL, Remarks on synonyms
of european spiders. Upsala &c. 1870–1873.
pp.[vii].645. [4000.]

THE ZOOLOGICAL RECORD [XI; *afterwards:* X; XI;
XI b; XII]. Arachnida [*afterwards:* including tardigrada and pentostomida; *afterwards:* to which are
added gigantostraca (xiphosura, trilobita, eurypterida), pantopoda, tardigrada, pentostomida,
myriopoda and prototracheata; *afterwards:* to
which are added arthropod groups other than
insecta and crustacea].

 xiii. 1876. By O[ctavius] P[ickard-]Cam-
  bridge. pp.20. [100.]
 xiv. 1877. pp.24. [100.]
 xv. [*not published.*]
 xvi. 1878–1879. pp.56. [200.]
 xvii. 1880. pp.30. [100.]
 xviii. 1881. pp.32. [100.]
 xix. 1882. pp.33. [100.]

Animals

xx. [*not published.*]

xxi. 1883–1884. By T. D. Gibson-Carmichael. pp.11. [129.]

xxii. 1885. By P[hilipp] Bertkau. pp.35. [200.]

xxiii. 1886. By R[eginald] Innes Pocock. pp.18. [97.]

xxiv. 1887. pp.32. [150.]

xxv. 1888. pp.28. [200.]

xxvi. 1889. pp.27. [100.]

xxvii. 1890. pp.20. [50.]

xxviii. 1891. pp.26. [100.]

xxix. 1892. pp.39. [150.]

xxx. 1893. pp.33. [150.]

xxxi. [*not published.*]

xxxii. 1894–1895. pp.56. [300.]

xxxiii. 1896. By Albert William Brown. pp.19. [150.]

xxxiv. 1897. pp.50. [151.]

xxxv. 1898. pp.39. [122.]

xxxvi. 1899. pp.31. [110.]

xxxvii. 1900. pp.38. [131.]

xxxviii. 1901. pp.47. [217.]

xxxix. 1902. By W. T. Calman. pp.63. [225.]

xl. 1903. pp.47. [190.]

xli. 1904. By Eugène Simon. pp.49. [236.]

xlii. 1905. pp.39. [197.]

xliii. 1906. By R[obert] Shelford. pp.55. [300.]

xliv. 1907. pp.55. [300.]

xlv. 1908. pp.52. [300.]

xlvi. 1909. By W. T. Calman. pp.56. [300.]

xlvii. 1910. pp.55. [300.]

xlviii. 1911. pp.53. [300.]

xlix. 1912. pp.47. [300.]

l. 1913. pp.54. [300.]

li. 1914. pp.34. [200.]

lii. 1915. pp.37. [200.]

liii. 1916. pp.32. [200.]

liv. 1917. pp.24. [125.]

lv. 1918. pp.22. [100.]

lvi. 1919. pp.26. [100.]

lvii. 1920. By Stephen K. Montgomery. pp.38. [200.]

lviii. 1921. pp.30. [150.]

lix. 1922. pp.36. [270.]

lx. 1923. By W[illiam] L[utley] Sclater. pp.31. [180.]

lxi. 1924. pp.39. [247.]

lxii. 1925. pp.42. [301.]

lxiii. 1926. pp.35. [298.]

lxiv. 1927. By Susan Finnegan. pp.52. [345.]

lxv. 1928. pp.51. [416.]

lxvi. 1929. pp.42. [338.]

lxvii. 1930. pp.51. [382.]

lxviii. 1931. pp.54. [424.]

lxix. 1932. pp.53. [380.]
lxx. 1933. pp.56. [471.]
lxxi. 1934. pp.51. [435.]
lxxii. 1935. pp.65. [494.]
lxxiii. 1936. By R. J. Whittick. pp.82. [554.]
lxxiv. 1937. pp.73. [541.]
lxxv. 1938. pp.33–88. [435.]
lxxvi. 1939. pp.46. [372.]
lxxvii. 1940. pp.36. [199.]
lxxviii. 1941. By E[rnest] Browning. pp.58.
  [425.]
lxxix. 1942. pp.70. [422.]
lxxx. 1943. pp.61. [426.]
lxxxi. 1944. pp.83. [564.]
lxxxii. 1945. pp.122. [956.]
lxxxiii. 1946. pp.84. [610.]
lxxxiv. 1947. pp.86. [596.]
lxxxv. 1948. pp.92. [645.]
lxxxvi. 1949. pp.98. [691.]
lxxxvii. 1950. pp.71. [648.]
lxxxviii. 1951. pp.105. [666.]
lxxxix. 1952.
xc. 1953. pp.110. [872.]
xci. 1954. pp.113. [830.]
xcii. 1955. pp.113. [813.]
xciii. 1956. pp.126. [1000.]
xciv. 1957. pp.116. [1100.]

xcv. 1958. pp.123. [1100.]
xcvi. 1959. pp.74. [1200.]
xcvii. 1960. pp.73. [1100.]
xcviii. 1961. pp.92. [1300.]

in progress; the earlier issues were not published separately; the issues for 1906–1914 also form part of vols.vi–xiv of the International catalogue of scientific literature, N. Zoology.

CARLO PALAU, Bibliografia aracnologica italiana. Pisa 1878. pp.15. [150.]

MELVILLE H. HATCH, A bibliographical catalogue of the injurious arachnids and insects of Washington. University of Washington: Publications in biology (vol.i, no.4): Seattle 1938. pp.163–223. [1000.]

PIERRE [NUMA LOUIS] BONNET, Bibliographia araneorum. Analyse méthodique de toute la littérature aranéologique jusqu'en 1939. Toulouse 1945–1961. pp.xviii.832+5058+591. [8000.]

Arthropoda.

ZOOLOGISCHER jahresbericht . . . [1880–1885: II. abtheilung:] Arthropoda. Zoologische Station zu Neapel: Leipzig.

1880. Redigirt von J. Vict[or] Carus. pp.iv.436. [1500.]

Animals

1881. Redigirt von Paul Mayer. pp.viii.420.
[1500.]

1882. Redigirt von P. Mayer und Wilh[elm]
Giesbrecht. pp.[iii].478. [1750.]

1883. pp.[iii].586. [2000.]

1884. pp.[iii].586. [2500.]

1885. pp.[iii].620. [2500.]

1886. Referenten: W. Giesbrecht, P. Mayer.
pp.82. [300.]

1887. pp.57. [250.]

1888. pp.75. [350.]

1889. pp.89. [600.]

1890. pp.74. [400.]

1891. pp.72. [450.]

1892. pp.90. [450.]

1893. pp.83. [400.]

1894. pp.73. [400.]

1895. pp.83. [400.]

1896. pp.67. [400.]

1897. pp.52. [400.]

1898. pp.57. [400.]

1899. pp.60. [400.]

1900. pp.56. [500.]

1901. pp.67. [400.]

1902. pp.72. [450.]

1903. pp.74. [450.]

1904. pp.78. [500.]

1905. pp.71. [550.]
1906. pp.77. [500.]
1907. pp.71. [550.]
1908. pp.80. [600.]
1909. pp.99. [650.]
1910. pp.75. [600.]
1911. pp.77. [650.]
1912. pp.85. [650.]

earlier issues and a latter issue were not published separately.

ERICH WASMANN, Kritisches verzeichniss der myrmekophilen und termitophilen arthropoden. Berlin 1894. pp.xv.232. [2000.]

THE ZOOLOGICAL RECORD. IX. Arthropoda.
xliii. 1906. By D[avid] Sharp. pp.3. [20.]
xliv. 1907. pp.2. [12.]
xlv. 1908. pp.3. [12.]
xlvi. 1909. pp.2. [10.]
xlvii. 1910. pp.2. [11.]
xlviii. 1911. pp.2. [5.]
xlix. 1912. pp.2. [8.]
l. 1913. pp.2. [12.]
li. 1914. pp.2. [3.]
lii. 1915. single leaf. [none.]
liii.1916. single leaf. [6.]
liv. 1917. single leaf. [3.]

lv. 1918. single leaf. [4.]

lvi. 1919. single leaf. [7.]

the earlier and later issues were not published as separate sections; the issues for 1906–1914 also form part of vols.vi–xiv of the International catalogue of scientific literature, N. Zoology.

HORMONES and their functions in arthropods ... 1953–1959. Science library: Bibliographical series (no.774): [1960]. pp.19. [375.]*

Aves. *see* **Ornithology.**

Batrachia. *see* **Reptilia.**

Bats.

NATURAL history and reproduction of British and european bat species from 1912–1938. Science library: Bibliographical series (no.412): 1938. ff.9. [175.]*

Bear.

JAMES R. TIGNER and DOUGLAS L. GILBERT, A contribution toward a bibliography on the black bear. Colorado department of game and fish: Technical bulletin (no.5): [Fort Collins] 1960. pp.[iii].42. [600.]*

Beaver.

LEE E[MMETT] YEAGER and KEITH G. HAY, A contribution toward a bibliography on the beaver. Colorado department of game and fish: Technical bulletin (no.1): [Fort Collins] 1955. pp.[ii]. ii.100. [1500.]*

Bee. *see* **Apiculture.**

Beetle.

BIONOMICS of beetles of the family tenebrionidae which occur in stored products. Science library: Bibliographical series (no.773): [1960]. pp.5. [100.]*

Birds. *see* **Ornithology.**

Bombyx mori.

[PIERRE AUGUSTIN BOISSIER DE SAUVAGES DE LA CROIX], Catalogue des auteurs qui ont écrit sur les vers à soie & sur les mûriers. [Nîmes 1763]. pp.8. [100.]

ROBERTO DI TOCCO, Bibliografia del filugello (Bombyx mori L.) e del gelso (Morus alba L.). Ente nazionale serico: Milano 1927. pp.xliii.262. [3850.]

LIOU-HO and CHARLES ROUX, Aperçu bibliographique sur les anciens traités chinois de botanique, d'agriculture, de sériculture et de fungiculture. Lyon 1927. pp.39. [75.]

REFERENCES on the silkworm, Bombyx mori, covering the period 1930–1934. Science library: Bibliographical series (no.162): 1934. ff.[i].10. [172.]★

Brachiopoda.

THE ZOOLOGICAL RECORD. [VIII; *afterwards*: VII]. Brachiopoda [1906–1936: and bryozoa].

> xxi. 1884. By Edward [Carl Eduard] von Martens. pp.2. [10.]
>
> xxii. 1885. By William E. Hoyle. pp.4. [13.]
>
> xxiii. 1886. pp.6. [25.]
>
> xxiv. 1887. pp.5. [25.]
>
> xxv. 1888. pp.6. [40.]
>
> xxvi. 1889. By Oswald H[awkins] Latter. pp.6. [22.]
>
> xxvii. 1890. By [sir] P[eter] Chalmers Mitchell. pp.9. [25.]
>
> xxviii. 1891. By B[ernard] B[arham] Woodward. pp.7. [36.]
>
> xxix. 1892. pp.8. [31.]
>
> xxx. 1893. pp.6. [13.]
>
> xxxi. 1894. pp.4. [17.]
>
> xxxii. 1895. pp.6. [27.]
>
> xxxiii. 1896. pp.4. [20.]
>
> xxxiv. 1897. By E. R. Sykes. pp.5. [41.]
>
> xxxv. 1898. pp.5. [39.]
>
> xxxvi. 1899. pp.7. [52.]

xxxvii. 1900. pp.6. [48.]

xxxviii. 1901. pp.12. [98.]

xxxix. 1902. pp.7 [58.]

xl. 1903. pp.6. [50.]

xli. 1904. pp.7. [60.]

xlii. 1905. pp.8. [58.]

xliii. 1906. By A. D. Imms, S. Pace and
   R. M. Pace. pp.28. [200.]

xliv. 1907. pp.39. [222.]

xlv. 1908. pp.28. [169.]

xlvi. 1909. By F. W. Edwards. pp.26. [150.]

xlvii. 1910. pp.16. [104.]

xlviii. 1911. pp.20. [102.]

xlix. 1912. pp.13. [56.]

l. 1913. pp.15. [65.]

li. 1914. pp.14. [85.]

lii. 1915. pp.14. [95.]

liii. 1916. pp.12. [52.]

liv. 1917. pp.15. [72.]

lv. 1918. pp.11. [74.]

lvi. 1919. pp.11. [65.]

lvii. 1920. By W. N. Edwards. pp.15. [69.]

lviii. 1921. pp.13. [67.]

lix. 1922. By H. M. Muir Wood. pp.16. [134.]

lx. 1923. pp.15. [132.]

lxi. 1924. pp.20. [148.]

lxii. 1925. pp.26. [218.]

lxiii. 1926. pp.36. [250.]

lxiv. 1927. pp.37. [214.]

lxv. 1928. By H. M. M. Wood [and Anna B. Hastings]. pp.42. [247.]

lxvi. 1929. pp.35. [213.]

lxvii. 1930. pp.42. [235.]

lxviii. 1931. pp.38. [205.]

lxix. 1932. pp.37. [222.]

lxx. 1933. pp.32. [206.]

lxxi. 1934. pp.37. [211.]

lxxii. 1935. pp.47. [347.]

lxxiii. 1936. By J. K. S. St.-Joseph [and A. B. Hastings]. pp.38. [275.]

lxxiv. 1937. By J. K. S. St.-Joseph. pp.38. [215.]

lxxv. 1938. pp.43. [161.]

lxxvi. 1939. pp.24. [146.]

lxxvii. 1940. pp.13. [77.]

lxxviii. 1941. pp.13. [65.]

lxxix. 1942. pp.17. [63.]

lxxx. 1943. pp.11. [51.]

lxxxi. 1944. pp.9. [50.]

lxxxii. 1945. pp.17. [119.]

lxxxiii. 1946. pp.15. [92.]

lxxxiv. 1947. pp.11. [72.]

lxxxv. 1948. By H. M. M. Wood. pp.33. [187.]

lxxxvi. 1949. pp.31. [195.]

lxxxvii. 1950. pp.28. [176.]

lxxxviii. 1951. pp.31. [211.]

lxxxix. 1952.

xc. 1953. pp.30. [199.]

xci. 1954. pp.28. [140.]

xcii. 1955. Compiled by D. V. Ager and E. F. Owen. pp.24. [182.]

xciii. 1956. pp.34. [200.]

xciv. 1957. Compiled by E. F. Owen. pp.24. [175.]

xcv. 1958. pp.18. [150.]

xcvi. 1959. pp.16. [150.]

xcvii. 1960. pp.20. [200.]

xcviii. 1961. pp.23. [200.]

in progress; earlier issues were not published separately; the issues for 1906–1914 also form part of vols.vi–xiv of the International catalogue of scientific literature, N. Zoology; *from vol.lxxiv bryozoa form a separate section.*

THOMAS DAVIDSON and W. H. DALTON, Bibliography of the brachiopoda. Palæontographical society: A monograph of the british fossil brachiopoda (vol.vi): 1886. pp.[iv].163. [3000.]

ZOOLOGISCHER jahresbericht . . . Brachiopoda. Zoologische station zu Neapel: Berlin.

1886. Referent: W. J. Vigelius. pp.2. [2.]
1887. pp.[2]. [3.]
amalgamated with the section for Bryozoa.

CHARLES SCHUCHERT, Synopsis of american fossil brachiopoda, including bibliography and synonymy. Geological survey: Bulletin (no.87): Washington 1897. pp.464. [6000.]

C. DIENER, Brachiopoda triadica. Fossilium catalogus (1.10): Berlin 1920. pp.109. [2500.]

CHARLES SCHUCHERT and CLARA M. LE VENE, Brachiopoda. [Generum et genotyporum index et bibliographia). Fossilium catalogus (1.42): Berlin 1929. pp.140. [2500.]

Bryozoa.

THE ZOOLOGICAL RECORD [IX; *afterwards:* VIII]. Polyzoa [*afterwards:* Bryozoa (polyzoa)].

xxi. 1884. By Edward von Martens. pp.15. [50.]
xxii. 1885. By George Robert Vine. pp.10. [24.]
xxiii. 1886. By William E. Hoyle. pp.12. [30.]
xxiv. 1887. pp.17. [50.]
xxv. 1888. pp.11. [50.]
xxvi. 1889. By Oswald H[awkins] Latter. pp.8. [40.]

xxvii. 1890. By [Sir] P[eter] Chalmers Mitchell. pp.10. [50.]

xxviii. 1891. By B[ernard] B[arham] Woodward. pp.7. [21.]

xxix. 1892. pp.5. [22.]

xxx. 1893. pp.7. [16.]

xxxi. 1894. pp.6. [20.]

xxxii. 1895. pp.6. [25.]

xxxiii. 1896. By Florence Buchanan. pp.6. [21.]

xxxiv. 1897. pp.4. [18.]

xxxv. 1898. pp.5. [20.]

xxxvi. 1899. By Arthur Willey. pp.6. [34.]

xxxvii. 1900. pp.8. [30.]

xxxviii. 1901. By Alice L. Embleton. pp.9. [59.]

xxxix. 1902. pp.10. [62.]

xl. 1903. pp.16. [102.]

xli. 1904. pp.17. [122.]

xlii. 1905. By Helen P. Kemp. pp.10. [83.]

lxxiv. 1937. By Anna B. Hastings. pp.9. [74.]

lxxv. 1938. pp.10. [72.]

lxxvi. 1939. pp.6. [37.]

lxxvii. 1940. pp.5. [36.]

lxxviii. 1941. pp.4. [33.]

lxxix. 1942. pp.7. [27.]

lxxx. 1943. pp.7. [44.]

lxxxi. 1944. pp.10. [61.]
lxxxii. 1945. pp.19. [86.]
lxxxiii. 1946. pp.4. [31.]
lxxxiv. 1947. pp.8. [38.]
lxxxv. 1948. pp.8. [61.]
lxxxvi. 1949. pp.15. [50.]
lxxxvii. 1950. pp.12. [76.]
lxxxviii. 1951. pp.7. [56.]
lxxxix. 1952.

xc. 1953. pp.12. [61.]
xci. 1954. pp.10. [80.]
xcii. 1955. pp.7. [49.]
xciii. 1956. pp.18. [90.]
xciv. 1957. pp.13. [75.]
xcv. 1958. pp.25. [125.]
xcvi. 1959. pp.11. [200.]
xcvii. 1960. pp.12. [200.]

*in progress; the earlier issues were not published
separately; the issues for 1906–1936 were amalgamated
with the section on brachiopoda.*

ZOOLOGISCHER jahresbericht.... Bryozoa [1888–
1912: und Brachiopoda]. Zoologische station zu
Neapel: Berlin.

1886. Referent: W. J. Vigelius. pp.8. [12.]
1887. pp.5. [12.]
1888. Referent: J. F. van Bemmelen. pp.6. [20.]
1889. pp.7. [20.]

1890. Referent: P. Schiemenz. pp.11. [20.]
1891. pp.12. [20.]
1892. pp.10. [20.]
1893. pp.9. [20.]
1894. pp.2. [4.]
1895. pp.2 [7].
1896. Referent: P. Mayer. pp.3. [7.]
1897. pp.4. [4.]
1898. pp.3. [5.]
1899. pp.2. [3.]
1900. pp.5. [8.]
1901. pp.2. [3.]
1902. pp.7. [8.]
1903. pp.2. [4.]
1904. pp.2. [7.]
1905. pp.3. [7.]
1906. pp.5. [10.]
1907. pp.2. [7.]
1908. pp.3. [10.]
1909. pp.2. [5.]
1910. pp.2. [3.]
1911. pp.2. [8.]
1912. pp.2. [11.]

earlier issues and a later issue were not published separately.

E. C. JELLY, A synonymic catalogue of the recent marine bryozoa. Including fossil synonyms. 1889.

pp.iii–xv.322. [9000.]

JOHN M. NICKLES and RAY S. BASSLER, A synopsis of american fossil bryozoa, including bibliography and synonymy. Geological survey: Bulletin (no.173): Washington 1900. pp.663. [6000.]

R. S. BASSLER, Bryozoa. (Generum et genotyporum index et bibliographia). Fossilium catalogus (1.67): 's Gravenhage 1934. pp.229. [4000.]

Buprestidae.

CH[ARLES] KERREMANS, Catalogue synonymique des buprestides décrits de 1758 à 1890. Société entomologique de Belgique: Mémoires (vol.i): Bruxelles 1892. pp.304.

Cats.

PERCY L. BABINGTON, A collection of books about cats. Cambridge 1918. pp.[iii].20. [75.]
54 copies printed.

SIDNEY DENHAM, Cats between covers. A bibliography of books about cats. 1952. pp.44. [167.]

Cephalopoda.

C. DIENER, Cephalopoda triadica. Fossilium catalogus (1. 8): Berlin 1915. pp.369. [6000.]
— — [Supplement. By] A. Kutassy . . . (1. 56): 1932. pp.iv.371–832. [4000.]

E. VON BÜLOW-TRUMMER, Cephalopoda dibranchiata. Fossilium catalogus (I. 11): Berlin 1920. pp.313. [6000.]

E[DWARD] M[ARTIN] KINDLE and A[RTHUR] M. MILLER, Bibliographic index of north american devonian cephalopoda. Geological society of America: Special papers (no.23): [Washington] 1939. pp.xi.179. [2500.]

Cerambycidae.

MAURICE PIC, Catalogue bibliographique et synonymique d'Europe et des régions avoisinantes. Matériaux pour servir à l'étude des longicornes (3ᵐᵉ cahier [2ᵉ partie]): [Lyon 1900]. pp.74. [2000.]

Cercopoidea.

Z[ENO] P[AYNE] METCALF, A bibliography of the cercopoidea (homoptera: auchenorhyncha). North Carolina state college: Paper (no.1135): [Raleigh] 1960. pp.iv.262. [3000.]
this is a partial new edition of a bibliography entered under Homoptera, below.

Cestoda.

CH[ARLES] WARDELL STILES and ALBERT HASSALL, Index-catalogue of medical and veterinary zoology. Subjects: cestoda and cestodaria. Hygienic laboratory: Bulletin (no.85): Washington 1912. pp.467. [30,000.]

Charadriiformes.

GEORGE CARMICHAEL LOW, The literature of the charadriiformes from 1894–1924. 1924. pp.xi.220. [2000.]

—— Second edition. . . . 1894–1928. 1931. pp. xiii.637. [7770.]

Chrysops.

REFERENCES to the early stages, larval habits and life-history of species of chrysops [corizoneura, corcyra cephalonica]. Science library: Bibliographical series (no.50): [1932]. ff.[i].4. [27.]*

Cicadidae.

W[ILLIAM] L[UCAS] DISTANT, A synonymic catalogue of homoptera. Part 1. Cicadidæ. British museum: 1906. pp.[iii].207. [4000.]

Z[ENO] P[AYNE] METCALF, A bibliography of cicadoidea (homoptera: auchenorhyncha). North Carolina state college: North Carolina agricultural experiment station: Paper (no.1373): Raleigh 1962. pp.iv.229. [2250.]

Cicindelidae.

[HERMANN WILHELM] WALTER HORN, Les cicindélides de Madagascar. Première partie. Catalogue bibliographique et synonymique. Académie malgache: Mémoire [part xx]: Tananarive 1934. pp. 3–28. [250.]

Clam.

H[AYES] T. PFITZENMEYER and CARL N[ATHANIEL] SHUSTER, A partial bibliography of the soft shell clam, *Mya arenaria* L. Maryland department of research and education: Chesapeake biological laboratory: Contribution (no.123): [Solomons] 1960. pp.29. [800.]*

Cnidaria.

C. DIENER, Cnidaria triadica. Fossilium catalogus (1. 13): Berlin 1921. pp.46. [1000.]

Coelenterata.

THE ZOOLOGICAL RECORD [XVI; *afterwards:* IV]. Cœlenterata.

    xiii. 1876. By C[hristian] F[rederik] Lütken. pp.16. [100.]

    xiv. 1877. pp.18. [100.]

    xv. 1878. pp.21. [100.]

    xvi. 1879. By Alfred Gibbs Bourne and Sydney J[ohn] Hickson. pp.21. [100.]

    xvii. 1880. pp.23. [100.]

    xviii. 1881. pp.14. [100.]

xix. 1882. pp.20. [100.]

xx. 1883. pp.16. [100.]

xxi. 1884. pp.15. [93.]

xxii. 1885. pp.10. [67.]

xxiii. 1886. By J[ohn] Arthur Thomson. pp. 14. [75.]

xxiv. 1887. pp.18. [112.]

xxv. 1888. By William E. Hoyle. pp.34. [150.]

xxvi. 1889. pp.24. [95.]

xxvii. 1890. By Sidney J[ohn] Hickson. pp.22. [96.]

xxviii. 1891. pp.14 [82.]

xxix. 1892. pp.13. [49.]

xxx. 1893. pp.16. [85.]

xxxi. [*not published*].

xxxii. 1894–1895. By R. T. Günther. pp.31. [218.]

xxxiii. 1896. pp.22. [122.]

xxxiv. [*not published*].

xxxv. 1897–1898. pp.44. [252.]

xxxvi. 1899. pp.23. [138.]

xxxvii. 1900. pp.24. [106.]

xxxviii. 1901. By Alice L. Embleton. pp.22. [153.]

xxxix. 1902. pp.31. [187.]

xl. 1903. pp.32. [234.]

xli. 1904. pp.35. [240.]

Animals

xlii. 1905. By Cora B. Sanders. pp.39. [187.]

xliii. 1906. By H. M. Woodcock and A. D. Imms. pp.27. [150.]

xliv. 1907. By H. M. Woodcock. pp.28. [180.]

xlv. 1908. pp.30. [199.]

xlvi. 1909. pp.27. [194.]

xlvii. 1910. B. F. W. Edwards. pp.27. [916.]

xlviii. 1911. By C. L. Boulenger. pp.20. [142.]

xlix. 1912. pp.15. [131.]

l. 1913. By A. Knyvett Totton. pp.24. [144.]

li. 1914. By F. W. Edwards. pp.12. [117.]

lii. 1915. pp.14. [116.]

liii. 1916. pp.9.]83.]

liv. 1917. pp.10. [104.]

lv. 1918. By A. K. Totton. pp.15. [89.]

lvi. 1919. pp.16. [93.]

lvii. 1920. pp.9. [67.]

lviii. 1921. pp.14. [101.]

lix. 1922. pp.20. [146.]

lx. 1923. pp.12. [117.]

lxi. 1924. pp.17. [144.]

lxii. 1925. pp.28. [206.]

lxiii. 1926. pp.23. [192.]

lxiv. 1927. pp.19. [155.]

lxv. 1928. pp.20. [168.]

lxvi. 1929. pp.19. [194.]

lxvii. 1930. pp.18. [178.]

lxviii. 1931. pp.29. [194.]

lxix. 1932. pp.26. [192.]

lxx. 1933. pp.23. [212.]

lxxi. 1934. pp.25. [184.]

lxxii. 1935. pp.24. [166.]

lxxiii. 1936. By H. Dighton Thomas. pp.43.
[231.]

lxxiv. 1937. pp.61. [294.]

lxxv. [*not published*].

lxxvi. [1938–]1939. pp.66. [500.]

lxxvii. 1940. pp.29. [117.]

lxxviii. 1941. pp.13. [93.]

lxxix. 1942. pp.20. [110.]

lxxx. 1943. pp.31. [111].

lxxxi. 1944. pp.23. [178.]

lxxxii. 1945. pp.35. [187.]

lxxxiii. 1946. pp.13. [95.]

lxxxiv. 1947. pp.47. [261.]

lxxxv. 1948. pp.30. [233.]

lxxxvi. 1949. pp.33. [168.]

lxxxvii. 1950. pp.54. [282.]

lxxxviii. 1951. pp.40. [274.]

lxxxix. 1952.

xc. 1953. pp.26. [218.]

xci. 1954. pp.40. [247.]

xcii. 1955. pp.44. [250.]

xciii. 1956. pp.43. [250.]

xciv. 1957. pp.63. [350.]

xcv. 1958. pp.62. [350.]

xcvi. 1959. pp.36. [300.]

xcvii. 1960. pp.34. [350.]

in progress; the earlier issues were not published separately; the issues for 1906–1914 also form vols.vi–xiv of the International catalogue of scientific literature, N. Zoology.

ZOOLOGISCHER jahresbericht. . . . Coelenterata. Zoologische station zu Neapel: Berlin.

  1886. Referenten: Paul Mayer, A. v. Heider.
    pp.23. [50.]

  1887. pp.24. [50.]

  1888. pp.33. [75.]

  1889. pp.28. [85.]

  1890. pp.29. [75.]

  1891. pp.25. [75.]

  1892. pp.20. [60.]

  1893. pp.19. [70.]

  1894. pp.18. [60.]

  1895. pp.18. [60.]

  1896. pp.19. [50.]

  1897. pp.22. [60.]

  1898. pp.22. [75.]

  1899. pp.18. [85.]

  1900. Referenten: P. Mayer, J[ames] H[artley]

Ashworth. pp.20. [90.]

1901. Referenten: E. Hentschel, J. H. Ashworth. pp.16. [65.]

1902. pp.28. [100.]

1903. Referenten: O[tto] Maas, J. H. Ashworth. pp.25. [125.]

1904. pp.41. [150.]

1905. pp.31. [150.]

1906. pp.37. [175.]

1907. pp.35. [125.]

1908. pp.34. [150.]

1909. pp.56. [200.]

1910. Referenten: E. Hentschel, J. H. Ashworth. pp.40. [175.]

1911. pp.34. [150.]

1912. pp.31. [150.]

earlier issues and a later issue were not published separately.

EUGÈNE LELOUP, Les cœlentérés de la faune belge. Leur bibliographie et leur distribution. Musée royal d'histoire naturelle de Belgique: Mémoires (no.107): Bruxelles 1947. pp.73. [500.]

Coleoptera.

L[OUIS JEAN RODOLPHE] AGASSIZ, Nomina systematica generum coleopterum. [Soloduri 1842–1846]. pp.xii.170. [6000.]

s[ILVIN] A[UGUSTIN] DE MARSEUL, Catalogue synonymique et géographique des coléoptères de l'ancien monde. [*c.*1890]. pp.iv.559. [35,000.]

[MAX] GEMMINGER and B. DE HAROLD, Catalogus coleopterorum huiusque descriptorum synonymicus et systematicus. Monachii 1868–1876. pp.xxxvi.424.[viii]. + [ii].425–752.[vi] + [ii].753–978.[vi] + [iii].979–1346.[viii] + [iii].1347–1608.[vi] + [ii].1609–1800.[vi] + [ii].1801–2180.[x] + [ii].2181–2668.[xii] + [ii].2669–2988.[xii] + [ii].2989–3232.[viii]. + [ii].3233–3478.[iv] + [ii].3479–3824.lxxiv. [175,000.]

supplements by several hands appear in the Mémoires *of the* Société royale des sciences de Liége (*1885*), *2nd ser. xi.*

MATHIAS RUPERTSBERGER, Biologie der käfer Europas. Eine übersicht der biologischen literatur. Linz a. d. Donau 1880. pp.xii.295. [5000.]
— — [supplement]. Die biologische literatur über die käfer Europas von 1880 an. Mit nachträgen aus früherer zeit. 1894. pp.viii.310. [5000.]

J[OSEPH] S[ANFORD] WADE, A contribution to a bibliography of the described immature stages of north american coleoptera. [Bureau of entomology and plant quarantine: Washington] 1935. ff.[i].114. [1500.]*

RICHARD E. BLACKWELDER, Checklist of the coleopterous insects of Mexico, central America, the West Indies and south America. National museum: Bulletin (no.185): Washington 1944–1946. pp.xii.763. [100,000.]

Conchology.

JOHANN SAMUEL SCHRÖTER, Für die litteratur und kenntniss der naturgeschichte, sonderlich der conchylien und der steine. Weimar 1782. pp. [xiv].278.[viii]+[xii].314.[viii]. [125.]

JOHANN SAMUEL SCHRÖTER, Neue litteratur und beyträge zur kenntniss der naturgeschichte, vorzüglich der conchylien und fossilien. Leipzig 1784–1787. pp.[viii].550.[xxx] + [viii.598.[xxxiv] + [vi].614.[xxvii]+[xii]456.[xxvi]. [500.]

LEWIS WESTON DILLWYN, A descriptive catalogue of recent shells. 1817. pp.xii.580+[vi].581–1092. [xxix]. [15,000.]

GEORGE W[ASHINGTON] TRYON, List of american writers on recent conchology. With the titles of their memoirs and dates of publications. New-York 1861. pp.68. [600.]

GEORGE W[ASHINGTON] TRYON, Publications of Isaac Lea on recent conchology. Philadelphia [printed] 1861. pp.13. [125.]

W[ILLIAM] G. BINNEY, Bibliography of north american conchology previous to the year 1860. Smithsonian institution: Miscellaneous collections (vols.v, ix): Washington 1863–1864. pp.vii.650+ v.306. [600.]

no more published.

Coral.

GILBERT RANSON and JACQUELINE PARETIAS, Coraux et récifs coralliens (bibliographie). Fondation Albert Ier: Institut océanographique: Bulletin (no.1121): Monaco 1958. pp.80. [1500.]

Corn borer.

J. S. WADE, A bibliography of the european corn borer (pyrausta nubilabis Hbn.). Department of agriculture: Miscellaneous circular (no.46): Washington 1925. pp.20. [450.]

— — [second edition]. 1928. pp.34. [800.]

Cotylosauria.

OSKAR KUHN, Cotylosauria et theromorpha. Fossilium catalogus (I. 79): 's-Gravenhage 1937. pp.209. [3000.]

Crinoidea.

WALTER BIESE [parts 77 &c.: and HERTHA SIEVERTS-DORECK], Crinoidea triadica [jurassica, cretacea, caenozoica]. Fossilium catalogus (1. 66; 70, 73, 76; 77; 80): Berlin [parts 70 &c.: s'-Gravenhage] 1934–1939. pp.255+[ii].379+254+151. [12,500.]

—— Supplementum . . . (1. 88): 1939. pp.81. [1000.]

Crocodilia.

OSKAR KUHN, Crocodilia. Fossilium catalogus (1. 75): 's-Gravenhage 1936. pp.144. [2000.]

Crustacea.

L[OUIS JEAN RODOLPHE] AGASSIZ, Nomina systematica generum crustaceorum. [Soloduri 1842–1846]. pp.vii.28.ii.10. [1300.]

THE ZOOLOGICAL RECORD [x; *afterwards:* IX; x]. Crustacea.

    xiii. 1876. By [Carl] Eduard von Martens. pp.18. [200.]

    xiv. 1877. pp.36. [300.]

    xv. 1878. pp.47. [400.]

    xvi. 1879. pp.45. [400.]

    xvii. 1880. pp.61. [500.]

    xviii. 1881. pp.38. [200.]

    xix. 1882. pp.39. [200.]

    xx. 1883. pp.34. [200.]

    xxi. 1884. pp.32. [200.]

xxii. 1885. By G. Herbert Fowler. pp.29.
[167.]

xxiii. 1886. pp.44. [223.]

xxiv. 1887. pp.26. [200.]

xxv. 1888. By Cecil Warburton. pp.30. [150.]

xxvi. 1889. pp.29. [150.]

xxvii. 1890. pp.23. [100.]

xxviii. 1891. pp.24. [100.]

xxix. 1892. By R[eginald] I[nnes] Pocock.
pp.34. [250.]

xxx. 1893. pp.36. [250.]

xxxi. [*not published.*]

xxxii. 1894–1895. pp.54. [400.]

xxxiii. 1896. By Florence Buchanan. pp.42.
[235.]

xxxiv. 1897. By Albert William Brown.
pp.43. [225.]

xxxv. 1898. pp.40. [167.]

xxxvi. 1899. pp.38. [157.]

xxxvii. 1900. pp.49. [192.]

xxxviii. 1901. pp.60. [316.]

xxxix. 1902. By W. T. Calman. pp.82. [334.]

xl. 1903. pp.66. [319.]

xli. 1904. pp.61. [292.]

xlii. 1905. pp.80. [385.]

xliii. 1906. By R. Shelford. pp.60. [350.]

xliv. 1907. pp.92. [382.]

xlv. 1908. pp.60. [396.]

xlvi. 1909. By W. T. Calman. pp.55. [337.]

xlvii. 1910. pp.47. [281.]

xlviii. 1911. pp.47. [312.]

xlix. 1912. pp.45. [317.]

l. 1913. pp.48. [330.]

li. 1914. pp.36. [245.]

lii. 1915. pp.26. [202.]

liii. 1916. pp.27. [170.]

liv. 1917. pp.21. [146.]

lv. 1918. pp.25. [136.]

lvi. 1919. pp.18. [119.]

lvii. 1920. By Stephen K. Montgomery. pp. 32. [218.]

lviii. 1921. pp.26. [186.]

lix. 1922. pp.29. [253.]

lx. 1923. By W. L. Sclater. pp.28. [219.]

lxi. 1924. pp.38. [338.]

lxii. 1925. pp.34. [302.]

lxiii. 1926. pp.45. [371.]

lxiv. 1927. pp.37. [378.]

lxv. 1928. pp.41. [390.]

lxvi. 1929. pp.42. [432.]

lxvii. 1930. pp.47. [445.]

lxviii. 1931. pp.52. [455.]

lxix. 1932. pp.49. [444.]

lxx. 1933. pp.53. [484.]

lxxi. 1934. pp.50. [507.]

lxxii. 1935. pp.61. [533.]

lxxiii. 1936. pp.66. [577.]

lxxiv. 1937. pp.55. [542.]

lxxv. 1938. pp.44. [434.]

lxxvi. 1939. pp.53. [481.]

lxxvii. 1940. pp.29. [246.]

lxxviii. 1941. pp.26. [225.]

lxxix. 1942. pp.33. [237.]

lxxx. 1943. pp.20. [159.]

lxxxi. 1944. By J. P. Harding. pp.22. [176.]

lxxxii. 1945. pp.44. [384.]

lxxxiii. 1946. pp.57. [498.]

lxxxiv. 1947. pp.40. [353.]

lxxxv. 1948. pp.51. [418.]

lxxxvi. 1949. pp.66. [635.]

lxxxvii. 1950. pp.68. [558.]

lxxxviii. 1951. pp.66. [556.]

lxxxix. 1952.

xc. 1953. pp.81. [796.]

xci. 1954. Compiled by J. P. Harding and R. W. Ingle. pp.89. [863.]

xcii. 1955. pp.95. [800.]

xciii. 1956. pp.77. [600.]

xciv. 1957. pp.99. [900.]

xcv. 1958. pp.110. [1000.]

xcvi. 1959. pp.72. [900.]

xcvii. 1960. pp.69. [900.]

in progress; earlier issues were not published separately; the issues for 1906–1914 also form part of vols.vi–xiv of the International catalogue of scientific literature, N. Zoology.

ANTHONY W[AYNE] VODGES, A bibliography of paleozoic crustacea from 1698 to 1889. Geological survey: Bulletin (no.63): Washington 1890. pp. 177. [3000.]

— — [second edition]. A classed and annotated bibliography of the palæozoic crustacea, 1698–1892. California academy of sciences: Occasional papers (vol.iv): San Francisco 1893. pp.[iv].413. [3500.]

a supplement by the author appears in the Proceedings *of the* California academy *(1895), 2nd ser. v.53–76.*

V. VAN STRAELEN, Crustacea eumalacostraca (crustaceis decapodis exclusis). Fossilium catalogus (1. 48): Berlin 1931. pp.98. [1250.]

Culicidae.

L. E. COOLING, A synonymic list of the more important species of culicidae of the australian region. Commonwealth of Australia: Department of health: Tropical division: Service publication (no.2): Melbourne [1924]. pp.61. [1000.]

Cypraeacea.

F. A. SCHILDER, Cypraeacea. Fossilium catalogus (1. 55): Berlin 1932. pp.276. [5000.]

Decapoda.

M. F. GLAESSNER, Crustacea decapoda. Fossilium catalogus (1. 41): Berlin 1929. pp.464. [7000.]

ROBERT GURNEY, Bibliography of the larvae of decapod crustacea: Ray society (no.125): 1939. pp.vii.123. [2500.]

a supplement appears in the author's Larvae of decapod crustacea *(1942), pp.288–299.*

Diatomaceae.

LIST of references on diatomaceous or infusorial earth. Library of Congress: Washington 1914. single leaf. [11.]*

FREDERICK W[ILLIAM] MILLS, An index to the general and species of the diatomaceae and their synonyms... 1816–1932. 1933–1934 [1935]. ff.[i]. 573+[i].574–1184+[i].1185–1726. [50,000.]*

Diptera.

L[OUIS JEAN RODOLPHE] AGASSIZ, Nomina systematica generum dipterorum. [Soloduri 1842–1846]. pp.vi.42. [1500.]

[BARON CARL] R[OBERT VON DEN] OSTEN SACKEN, Catalogue of the described diptera of north America. Smithsonian institution: Miscellaneous collections (vol.iii). Washington 1858. pp.xx.95. [5000.]

—— Second edition. . . . (no.270): 1878. pp. xlviii.276. [6000.]

J[OHN] M[ERTON] ALDRICH, A catalogue of north american diptera (or two-winged flies). Smithsonian institution: Miscellaneous collections (vol. xlvi): Washington 1905. pp.[ii].680. [25,000.]

T[RON] SOOT-RYEN, A review of the literature on norwegian diptera until the year 1940. Tromsø museum: Årshefter (vol.lxv, no.3): Tromsø 1943. pp.46. [100.]

Dog.

BIBLIOGRAPHIE für hundefreunde und jäger. Verzeichniss sämmtlicher schriften über hundezucht, dressur, krankheit etc. welche von 1840 bis 1879 im deutschen buchhandel erschienen sind, nebst einigen älteren guten werken. Gracklauer's fachkatalog (no.5): Leipzig 1879. pp.16. [125.]

MARCUS M. MASON, Bibliography of the dog. Ames, Iowa [1959]. pp.vii.401. [12,369.]

POLICE use of dogs. A selected bibliography. Los

Angeles police library bibliography series (no.2): Los Angeles 1959. ff.[i].4. [60.]★

Dogfish.

SOME references on hermaphroditism in the dogfish. Science library: Bibliographical series (no. 768): [1959]. single leaf. [15.]★

Donkey.

G. A. KNYAZEVA, Литература по биологии лошади, осла и мула, по коневодству и конному спорту, изданная в СССР в 1917–1961 гг. (Систематический указатель.) Московская . . . сельскохозяйственная академия имени К. Н. Тимирязева: Москва 1962. pp.654. [7500.]

Drosophila.

BIBLIOGRAPHY on the genetics of drosophila.

    i. By H[ermann] J[oseph] Muller. Imperial bureau of animal breeding and genetics: Edinburgh 1939. pp.132. [2965.]

    ii. By Irwin H[ermann] Herkowitz. Commonwealth bureau [&c.]: Farnham Royal [1952]. pp.xi.212.

    iii. By I. H. Herkowitz. Indiana university publications: Science series (no.21): Bloomington 1958. pp.ix.296. [3100.]★

iv. 1963. pp.[iii].344. [3305.]
in progress.

Earthworms.

EARTHWORMS. Public library of South Australia:
Research service: [Adelaide] 1959. ff.11. [185.]*

Echinoderma.

L[OUIS JEAN RODOLPHE] AGASSIZ, Nomina
systematica generum echinodermatum. [Soloduri
1842–1846]. pp.v.14.4. [600.]

THE ZOOLOGICAL record [XIV; *afterwards:* v].
Echinodermata [*afterwards:* Echinoderma].

    xiii. 1876. By C[hristian] F[rederik] Lütken.
        pp.16. [300.]

    xiv. 1877. pp.12. [250.]

    xv. 1878. pp.13. [250.]

    xvi. 1879. By F. Jeffrey Bell. pp.9. [200.]

    xvii. 1880. pp.11. [200.]

    xviii. 1881. pp.9. [200.]

    xix. 1882. pp.10. [200.]

    xx. 1883. pp.8. [200.]

    xxi. 1884. pp.10. [200.]

    xxii. 1885. pp.9. [53.]

    xxiii. 1886. pp.8. [59.]

    xxiv. 1887. pp.14. [150.]

    xxv. 1888. By Oswald H. Latter. pp.16. [64.]

xxvi. 1889. pp.2. [84.]

xxvii. 1890. By E. A. Minchin. pp.24. [100.]

xxviii. 1891. pp.91. [186.]

xxix. 1892. pp.22. [55.]

xxx. 1893. By F. A. Bather. pp.107. [288.]

xxxi. 1894. pp.55. [133.]

xxxii. 1895. pp.70. [181.]

xxxiii. [*not published*].

xxxiv. 1896–1897. pp.135. [358.]

xxxv. 1898. pp.73. [231.]

xxxvi. 1899. pp.101. [261.]

xxxvii. 1900. pp.153. [370.]

xxxviii. 1901. pp.99. [338.]

xxxix. 1902. pp.88. [306.]

xl. 1903. pp.105. [339.]

xli. 1904. pp.96. [300.]

xlii. 1905. By M. Grant. pp.92. [372.]

xliii. 1906. By A. D. Imms *and others*. pp.55. [250.]

xliv. 1907. By S. Pace and R. M. Pace. pp.40. [235.]

xlv. 1908. pp.30. [235.]

xlvi. 1909. By H. L. Hawkins. pp.54. [313.]

xlvii. 1910. pp.51. [377.]

xlviii. 1911. pp.48. [249.]

xlix. 1912. pp.61. [402.]

l. 1913. pp.50. [371.]

li. 1914. pp.29. [242.]
lii. 1915. By H. B. Preston. pp.31. [224.]
liii. 1916. pp.18. [168.]
liv. 1917. By M. Connolly. pp.24. [142.]
lv. 1918. By H. B. Preston. pp.24. [127.]
lvi. 1919. pp.16. [94.]
lvii. 1920. pp.20. [138.]
lviii. 1921. pp.20. [133.]
lix. 1922. pp.23. [186.]
lx. 1923. pp.23. [179.]
lxi. 1924. pp.30. [203.]
lxii. 1925. pp.28. [214.]
lxiii. 1926. pp.31. [287.]
lxiv. 1927. pp.31. [265.]
lxv. 1928. pp.34. [237.]
lxvi. 1929. pp.30. [259.]
lxvii. 1930. By A. G. Brighton. pp.30. [224.]
lxviii. 1931. pp.23. [183.]
lxix. 1932. pp.23. [189.]
lxx. 1933. pp.26. [168.]
lxxi. 1934. pp.24. [178.]
lxxii. 1935. pp.30. [256.]
lxxiii. 1936. pp.29. [183.]
lxxiv. 1937. pp.28. [205.]
lxxv. 1938. pp.30. [190.]
lxxvi. 1939. pp.24. [193.]
lxxvii. 1940. pp.18. [118.]

Animals

lxxviii. 1941. pp.12. [84.]
lxxix. 1942. pp.11. [80.]
lxxx. 1943. pp.10. [74.]
lxxxi. 1944. pp.27. [120.]
lxxxii. 1945. pp.14. [120.]
lxxxiii. 1946. pp.12. [101.]
lxxxiv. 1947. pp.11. [102.]
lxxxv. 1948. pp.17. [113.]
lxxxvi. 1949. pp.18. [161.]
lxxxvii. 1950. pp.19. [113.]
lxxxviii. 1951. pp.50. [344.]
lxxxix. 1952.
xc. 1953. pp.30. [245.]
xci. 1954. Compiled by A. M. Clark. pp.23. [188.]
xcii. 1955. pp.31. [227.]
xciii. 1956. pp.21. [200.]
xciv. 1957. pp.29. [250.]
xcv. 1958. pp.30. [275.]
xcvi. 1959. pp.18. [250.]
xcvii. 1960. pp.18. [250.]
xcviii. 1961. pp.33. [300.]

in progress; the earlier issues were not published separately; the issues for 1906–1914 also form part of vols.vi–xiv of the International catalogue of scientific literature, N. Zoology.

ZOOLOGISCHER jahresbericht. . . . Echinodermata

Animals

[*afterwards:* Echinoderma]. Zoologische station zu
Neapel: Berlin.

    1886. Referent: P. H. Carpenter. pp.12. [50.]

    1887. pp.17. [50.]

    1888. pp.26. [25.]

    1889. pp.23. [50.]

    1890–1891. Referent: Hubert Ludwig. pp.29.
    [150.]

    1892. pp.20. [75.]

    1893. pp.11. [65.]

    1894. pp.14. [65.]

    1895. pp.16. [75.]

    1896. pp.17. [90.]

    1897. pp.10. [65.]

    1898. pp.14. [75.]

    1899. pp.12. [75.]

    1900. pp.15. [90.]

    1901–1902. pp.25. [150.]

    1903. pp.17. [125.]

    1904. pp.13. [100.]

    1905. pp.14. [100.]

    1906. pp.14. [100.]

    1907. pp.16. [100.]

    1908. pp.13. [100.]

    1909. pp.17. [125.]

    1910. pp.15. [125.]

    1911. pp.16. [100.]

1912. Referent: August Reichensperger. pp. 17. [150.]

earlier issues and a later issue were not published separately.

Echinoidea.

W. DEECKE, Echinoidea jurassica. Fossilium catalogus (I. 39): Berlin 1928. pp.540. [10,000.]

Eels. *see* Anguilla.

Epizoa.

WILLIAM FERDINAND ERICHSON, Nomina systematica generum epizoorum. [Soloduri 1842–1846]. pp.[ii].2. [30.]

forms part of the Nomenclator zoologicus *of L. J. R. Agassiz.*

Eurypterida.

C. DIENER, Eurypterida. Fossilium catalogus (I. 25): Berlin 1924. pp.[ii].29. [400.]

Fauna. *see* Zoology.

Ferret.

WERNER K. G. MOEBES, Bibliographie des kaninchens. Nebst anhang I. Das frettchen, II. Das meerschweinchen. Halle &c. [1946]. pp.xvi.318. [ferret: 20.]

Flea. *see* **Pulex.**

Foraminifera.

CHARLES DAVIES SHERBORN, A bibliography of the foraminifera, recent and fossil, from 1565–1888. 1888. pp.vii.152. [2250.]

A[DALBERT] LIEBUS and H[ANS] E[RNEST] THAL-MANN, Bibliographia foraminiferum recentium et fossilium. Fossilium catalogus (1. 59, 49, 60): Berlin 1931–1933. pp.179+36+28. [6000.]
the second part was published first.

JOSEPH AUGUSTINE CUSHMAN, A bibliography of american foraminifera. Cushman laboratory for foraminiferal research: Special publication (no.3): Sharon, Mass. 1932. pp.40. [650.]

BROOKS F[LEMING] ELLIS and ANGELINA R. MESSINA, Catalogue of foraminifera. Bibliography. American museum of national history: New York 1940. pp.[ii].270. [5400.]*

Frog.

FROG farming. Science library: Bibliographical series (no.750): [1957]. single leaf. [17.]*

Gastropoda.

W. WENZ, Gastropoda extramarina tertiaria. Fossilium catalogus (1. 17, 18, 20–23, 32, 38, 40, 43,

46): Berlin 1923–1930. pp.3387. [70,000.]

G[USTAV] HABER, Gastropoda, amphineura et scaphopoda jurassica. Fossilium catalogus (1. 53, 65 &c.): Berlin 1932 &c.
in progress.

SELECTED references on feeding mechanisms in prosobranch gastropods with special reference to littorina littorea, 1930–1944. Science library: Bibliographical series (no.643): 1947. single leaf. [12.]★

Glossophora.

C. DIENER, Glossophora triadica. Fossilium catalogus (1. 34): Berlin 1926. pp.[ii].242. [7500.]

Gryllidae.

CRICKETS, gryllidae, with special reference to their subterranean life and their destruction. References covering the period 1930–1934. Science library: Bibliographical series (no.202): 1935. ff.4. [59.]★

Hemiptera.

L[OUIS JEAN RODOLPHE] AGASSIZ, Nomina systematica generum hemipterorum. [Soloduri 1842–1846]. pp.viii.20.14. [1200.]

v[ALENTIN LVOVICH] BIANCHI, Enumeratio operum opusculorumque ad faunam hemipterorumheteropterorum imperii Rossici pertinentium, 1798–1897. St-Pétersbourg 1898. pp.36. [300.]

HOWARD MADISON PARSHLEY, A bibliography of the north american hemiptera-heteroptera. Smith college: Fiftieth anniversary publications (vol.iv): Northampton, Mass. 1925. pp.[ii].ix.252. [xxvii]. [2000.]

the unnumbered pages at the end of the volume are blank.

Herpetology. *see* Reptilia.

Hominidae.

WERNER and ANNEMARIE QUENSTEDT, Hominidae fossiles. Fossilium catalogus (I. 74): 's-Gravenhage 1936. pp.456. [12,500.]

Homoptera.

ZENO PAYNE METCALF, A bibliography of the homoptera (auchenorhyncha). University of North Carolina: North Carolina state college of agriculture and engineering: Department of zoology and entomology: Raleigh 1945. pp.[ii]. 886+3–186. [9000.]

a partial new edition is entered under Cercopoidea.

Animals

Hydrozoa.

O. KÜHN, Hydrozoa. Fossilium catalogus (1. 36): Berlin 1928. pp.[ii].114. [2000.]

Hymenoptera.

L[OUIS JEAN RODOLPHE] AGASSIZ, Nomina systematica generum hymenopterorum. [Soloduri 1842–1846]. pp.viii.36. [1300.]

ALEXANDER [SÁNDOR] MOCSÁRY, Literatura hymenoptera. Magyar nemzeti múzeum: Természetrajzi füzetek (vol.vi, parts i–iv): Budapest 1883. pp.122. [2000.]

Ichthyology. *see* **Fish.**

Ichthyosauria.

OSKAR KUHN, Ichthyosauria. Fossilium catalogus (1. 63): Berlin 1934. pp.75. [750.]

Inversicatenales.

O[ENE] POSTHUMUS, Inversicatenales. (Botryopterideae et zygopterideae). Fossilium catalogus (1. 12): Berlin 1926. pp.56. [1000.]

Invertebrata.

C. A. WHITE and H[ENRY] ALLEYNE NICHOLSON, Bibliography of north american invertebrate peleontology . . . including the West Indies and

Greenland. Geological survey: Miscellaneous publication (no.10): Washington 1878. pp.132. [1250.]

JOHN BELKNAP MARCOU, Bibliography of publications relating to the collection of fossil invertebrates in the United States national museum, including complete lists of the writings of Fielding B. Meek, Charles A. White, and Charles D. Walcott. United States national museum: Bulletin (no.30): Washington 1885. pp.333. [400.]

CORNELIUS BRECKINRIDGE BOYLE, A catalogue and bibliography of north american mesozoic invertebrata. Geological survey: Bulletin (no. 102): Washington 1893. pp.315. [6000.]

STUART WELLER, A bibliographic index of north american carboniferous invertebrates. Geological survey: Bulletin (no.153): Washington 1898. pp. 653. [12,000.]

FRANCIS LUTHER WHITNEY, Bibliography and index of north american mesozoix invertebrata. Bulletins of american paleontology (vol.xii. no.48): Ithaca, N.Y. 1928. pp.[ii].448. [5000.]

CARL C[OTTON] BRANSON, Bibliographic index of permian invertebrates. Geological society of America: Memoir (no.26): [Washington] 1948. pp.vii.1049. [17,500.]

Animals

Isoptera.

THOMAS E[LLIOTT] SNYDER, Catalog of the termites (isoptera) of the world. Smithsonian institution: Smithsonian miscellaneous collections (vol. 112): Washington 1949. pp.[ii].490. [6000.]

Ixodoidea.

GEORGE H[ENRY] F[ALKINER] NUTTALL, L. E. ROBINSON and W. F. COOPER, Bibliography of the ixodoidea. Ticks, a monograph of the ixodoidea: Cambridge 1911. pp.vi.ff.68. [1000.]

—— II. By G. H. F. Nuttall and L. E. Robinson. 1915. pp.[vi].ff.32. [462.]

Labyrinthodontia.

OSKAR KUHN, Labyrinthodontia. Fossilium catalogus (I. 61): Berlin 1933. pp.114. [1750.]

Lamellibranchiata.

W. TEPPNER, Lamellibranchiata tertiaria. 'Anisomyaria'. Fossilium catalogus (I. 2. 15): Berlin 1914, 1922. pp.296. [5000.]

C. DIENER, Lamellibranchiata triadica. Fossilium catalogus (I. 19): Berlin 1923. pp.260. [5000.]

—— [supplement by] A. Kutassy . . . (I. 51): 1931. pp.iv.261–477. [3000.]

345

Lepidoptera.

L[OUIS JEAN RODOLPHE] AGASSIZ, Nomina systematica generum lepidopterorum. [Soloduri 1842–1846]. pp.vi.70. [2500.]

JOHN G[OTTLIEB] MORRIS, Catalogue of the described lepidoptera of north America. Smithsonian institution: Miscellaneous collections (vol. iii): Washington 1860. pp.viii.68. [2000.]

WILLIAM F[ORSELL] KIRBY, A synonymic catalogue of diurnal lepidoptera. 1871. pp.vii.690. [20,000.]

HENRY EDWARDS, Bibliographical catalogue of the described transformations of north american lepidoptera. National museum: Bulletin (no.35): Washington 1889. pp.147. [5000.]

W. H. MISKIN, A synonymical catalogue of the lepidoptera Rhopalocera (butterflies) of Australia, with full bibliographical reference. Queensland museum: Annals (no.1): Brisbane 1891. pp.[ii]. xx.93.ff.viii. [2500.]

HARRISON G. DYAR, A list of north american lepidoptera and key to the literature of this order of insects. National museum: Bulletin (no.52): Washington 1902. pp.xix.723. [20,000.]

LEPIDOPTERORUM catalogus. Editus a H[ans] Wagner. Berlin 1911 &c.

Animals

JAN ROMANISZYN and FREDERYK SCHILLE, Fauna motyli Polski. Polska akademja umiejętności: Komisja fizjograficzna: Prace monograficzne (vols.vi–vii): Krakowie 1929–1930[1931]. pp.555 +358. [2708.]

ORAZIO QUERCI, Bibliografia dei ropaloceri del Portogallo. Lisboa 1931–1932. pp.[ii].112. [250.]

W[ALTER] F[REDERICK] JEPSON, A critical review of the world literature on the lepidopterous stalk borers of tropical graminaceous crops. Commonwealth institute of entomology: 1954. pp.vi.128. [500.]

ARTHUR A[DRIAN] LISNEY, A bibliography of british lepidoptera 1608–1799. 1960. pp.[ii]. xviii.315. [419.]
privately printed.

Libellulae.

ALFRED PREUDHOMME DE BORRE, Répertoire alphabétique des noms spécifiques admis ou proposés dans la sous-famille des libellulines. Bruxelles 1889. pp.38. [500.]

Lizards.

THE ANATOMY, physiology and habits of lizards. Science library: Bibliographical series (no.69): [1933]. ff.[i].9. [125.]*

Lobster.

CHARLES E. DAWSON, A bibliography of the lobster and the spiny lobster, families homaridae and palinuridae. Florida state board of conservation: [Tallahassee] 1954. pp.[v].86. [1000.]

Locusts.

G. B. BUGDANOV, Русская литература по вредным саранчевым. Станция защиты растений при горском сельско-хозяйственном институте [no.13]: Владикавказ 1929. pp.49. [950.]

Malacology. *see* Mollusca.

Mammalia.

L[OUIS JEAN RODOLPHE] AGASSIZ, Nomina systematica generum mammalium. [Soloduri 1842–1846]. pp.xii.36.10. [1500.]

THE ZOOLOGICAL RECORD. [II; *afterwards:* XVIII; XVI; XVII; XVIII]. Mammalia.

    xiii. 1876. By Edward Richard Alston. pp.24. [300.]

    xiv. 1877. pp.24. [300.]

    xv. 1878. pp.24. [300.]

    xvi. 1879. By W. A. Forbes. pp.28. [300.]

    xvii. 1880. pp.31. [400.]

    xviii. 1881. pp.31. [400.]

Animals

xix. 1882. By Oldfield Thomas. pp.40. [500.]

xx. 1883. pp.59. [750.]

xxi. 1884. By W. L. Sclater. pp.52. [373.]

xxii. 1885. pp.48. [346.]

xxiii. 1886. pp.62. [406.]

xxiv. 1887. By R[ichard] Lydekker. pp.56.
   [400.]

xxv. 1888. pp.64. [500.]

xxvi. 1889. pp.55. [400.]

xxvii. 1890. pp.51. [400.]

xxviii. 1891. pp.58. [400.]

xxix. 1892. pp.55. [400.]

xxx. 1893. pp.42. [300.]

xxxi. 1894. pp.47. [300.]

xxxii. 1895. pp.40. [259.]

xxxiii. 1896. pp.33. [291.]

xxxiv. 1897. pp.36. [343.]

xxxv. 1898. pp.35. [330.]

xxxvi. 1899. pp.37. [338.]

xxxvii. 1900. pp.36. [346.]

xxxviii. 1901. pp.40. [361.]

xxxix. 1902. pp.44. [409.]

xl. 1903. pp.42. [385.]

xli. 1904. pp.51. [517.]

xlii. 1905. pp.47. [482.]

xliii. 1906. pp.83. [750.]

xliv. 1907. pp.74. [893.]

xlv. 1908. pp.84. [1028.]

xlvi. 1909. pp.75. [848.]

xlvii. 1910. pp.64. [706.]

xlviii. 1911. pp.51. [559.]

xlix. 1912. pp.54. [600.]

l. 1913. pp.54. [559.]

li. 1914. By R. Lydekker and Martin A. C. Hinton. pp.60. [776.]

lii. 1915. By M. A. C. Hinton. pp.48. [508.]

liii. 1916. pp.48. [542.]

liv. 1917. pp.28. [267.]

lv. 1918. pp.34. [300.]

lvi. 1919. pp.27. [235.]

lvii. 1920. pp.28. [286.]

lviii. 1921. pp.36. [459.]

lix. 1922. pp.60. [731.]

lx. 1923. By W. L. Sclater. pp.54. [691.]

lxi. 1924. pp.56. [679.]

lxii. 1925. pp.55. [731.]

lxiii. 1926. pp.66. [845.]

lxiv. 1927. pp.66. [910.]

lxv. 1928. pp.74. [1000.]

lxvi. 1929. pp.76. [967.]

lxvii. 1930. pp.73. [1015.]

lxviii. 1931. pp.77. [1054.]

lxix. 1932. pp.85. [1109.]

lxx. 1933. pp.83. [1152.]

lxxi. 1934. pp.85. [1212.]

lxxii. 1935. pp.93. [1235.]

lxxiii. 1936. pp.88. [1224.]

lxxiv. 1937. pp.98. [1341.]

lxxv. 1938. pp.91. [1307.]

lxxvi. 1939. pp.86. [1231.]

lxxvii. 1940. pp.54. [727.]

lxxviii. 1941. pp.67. [918.]

lxxix. 1942. pp.80. [1067.]

lxxx. 1943. By the late W. L. Sclater and F. C. Fraser. pp.63. [828.]

lxxxi. 1944. By F. C. Fraser. pp.40. [580.]

lxxxii. 1945. By R. W. Hayman. pp.81. [1130.]

lxxxiii. 1946. pp.98. [1357.]

lxxxiv. 1947. pp.104. [1370.]

lxxxv. 1948. pp.86. [1130.]

lxxxvi. 1949. pp.94. [1318.]

lxxxvii. 1950. pp.110. [1546.]

lxxxviii. 1951. pp.112. [1635.]

lxxxix. 1952.

xc. 1953. pp.214. [1835.]

xci. 1954. pp.124. [1796.]

xcii. 1955. pp.129. [1778.]

xciii. 1956. pp.107. [1500.]

xciv. 1957. pp.110. [1500.]

xcv. 1958. pp.125. [1500.]

xcvi. 1959. Compiled by J. E. Hill. pp.102. [2000.]

xcvii. 1960. pp.112. [2250.]
xcviii. 1961. pp.118. [2250.]
*in progress; earlier issues were not published
separately; the issues for 1906–1914 also form part
of vols.vi–xiv of the* International catalogue of
scientific literature, N. Zoology.

É[DOUARD] L[OUIS] TROUESSART, Catalogus
mammalium tam viventium quam fossilium. . . .
Nova editio. Berolini [1897–] 1898–1899. pp.vi.
iii–vi.664+v.665–1469. [40,000.]
the first edition appeared in journals, 1879–1886.

H. FISCHER-SIGWART, Säugethiere. Centralkommission für schweizerische landeskunde: Bibliographie der schweizerischen landeskunde (section
IV.6.3α): Bern 1900. pp.xv.104. [2000.]

CARLOS A[NTONIO] MARELLI, Bibliografia euris-
tica de los mamíferos, de caza y caza maritima.
La Plata 1936. pp.189. [1750.]

FREDERICK NUTTER CHASEN, A handlist of
malaysian mammals. Raffles museum: Bulletin
(no.15): Singapore 1940. pp.xx.209. [2000.]
—— [Sir] J[ohn] R[eeves] Ellerman and
T[erence] C[harles] S[tuart] Morrison Scott,
Supplement. . . . Containing a generic synonymy
and a complete index. 1955. pp.[ii].66. [500.]

Animals

RICHARD G. BEIDLEMAN, A partial bibliography of techniques in mamalogy. Rocky Mountain nature association: Estes Park 1951. ff.[i].i.76. [1000.]*

ANNIE P. GRAY, Mammalian hybrids. A checklist with bibliography. Commonwealth bureau of animal breeding and genetics: Technical communication (no.10): Farnham Royal 1954. pp.ix. 144. [600.]

ARTHUR TINDELL HOPWOOD and JUNE PAMELA HOLLYFIELD, An annotated bibliography of the fossil mammals of Africa (1742–1950). British museum (natural history): Fossil mammals of Africa (no.8): 1954. pp.[ii].194. [1000.]

Marine-borer.

WILLIAM F. CLAPP and ROMAN KENK, Marine borers. A preliminary bibliography. Library of Congress: Technical information division: Washington 1956–1957. pp.ix.346+viii.350. [1685.]*

— — [another edition]. Department of the navy: Office of naval research: 1963. pp.xii.1136. [2500.]*

Marsupialia.

J. J. FLETCHER, Catalogue of papers and works relating to the mammalian orders, marsupialia and monotremata. [Sydney 1885]. pp.55. [500.]

G. G. SIMPSON, Post-mesozoic marsupialia. Fossilium catalogus (I. 47): Berlin 1930. pp.87. [1250.]

Medusae.

A. KIESLINGER, Medusae fossiles. Fossilium catalogus (I. 26): Berlin 1924. pp.20. [300.]

Mesosauria.

OSKAR KUHN, Protorosauria. Mesosauria. Fossilium catalogus (I. 85): 's-Gravenhage 1939. pp. [ii].8.6. [Mesosauria: 25.]

Metatetranychus Ulmi.

JOAN R[OSALIND] GROVES, A synopsis of the world literature on the fruit tree red spider mite . . . and its predators. Commonwealth institute of entomology: 1951. pp.180. [1039.]

Mite.

JOSEF R[UDOLF] WINKLER, Bibliografie roztočů (*acari*) ČSR. Krajské vlastivědné museum: Praha 1959. pp.53. [661.]

Mocking-bird.

LIST of references on the mocking-bird. Library of Congress: Washington 1914. ff.2. [19.]*

Moles.

CONTROL and extermination of moles and rabbits. Science library: Bibliographical series (no.331): 1937. ff.6. [136.]*

Mollusca.

GIO[VANNI] BATTISTA BONOLA, Della bibliografia malacologica italiana. Milano 1839. pp.55. [100.]

L[OUIS JEAN RODOLPHE] AGASSIZ, Nomina systematica generum molluscorum. [Soloduri 1842–1846]. pp.xiv.98. [3500.]

THE ZOOLOGICAL RECORD. [VII; *afterwards:* VIII; IX]. Mollusca.

    xiii. 1876. By Eduard von Martens. pp.67.
       [1000.]
    xiv. 1877. pp.98. [1000.]
    xv. 1878. pp.95. [1000.]
    xvi. 1879. pp.113. [1000.]
    xvii. 1880. pp.125. [1000.]
    xviii. 1881. pp.108. [1000.]
    xix. 1882. pp.115. [1000.]
    xx. 1883. pp.113. [1000.]
    xxi. 1884. pp.86. [1000.]
    xxii. 1885. By W. A. Hoyle. pp.106. [1000.]
    xxiii. 1886. pp.102. [1000.]
    xiv. 1887. pp.85. [500.]

xxv. 1888. pp.82. [500.]

xvi. 1889. By [sir] P[eter] Chalmers Mitchell. pp.85. [500.]

xxvii. 1890. pp.71. [400.]

xxviii. 1891. By B[ernard] B[arham] Woodward. pp.114. [479.]

xxix. 1892. pp.96. [441.]

xxx. 1893. pp.77. [419.]

xxxi. 1894. pp.87. [445.]

xxxii. 1895. pp.83. [491.]

xxxiii. 1896. pp.59. [391.]

xxxiv. 1897. By E. R. Sykes. pp.78. [527.]

xxxv. 1898. pp.79. [531.]

xxxvi. 1899. pp.87. [595.]

xxxvii. 1900. pp.92. [588.]

xxxviii. 1901. pp.102. [675.]

xxxix. 1902. pp.85. [611.]

xl. 1903. pp.85. [588.]

xli. 1904. pp.92. [670.]

xlii. 1905. pp.86. [653.]

xliii. 1906. By E. R. Sykes, S. Pace and R. M. Pace. pp.103. [650.]

xliv. 1907. pp.179. [767.]

xlv. 1908. pp.128. [932.]

xlvi. 1909. By H. B. Preston. pp.95. [710.]

xlvii. 1910. pp.96. [798.]

xlviii. 1911. pp.103. [732.]

xlix. 1912. pp.102. [700.]

Animals

lxxvi. 1939. pp.151. [1056.]
lxxvii. 1940. pp.68. [486.]
lxxviii. 1941. pp.70. [364.]
lxxix. 1942. pp.64. [444.]
lxxx. 1943. pp.55. [396.]
lxxxi. 1944. pp.59. [427.]
lxxxii. 1945. pp.101. [902.]
lxxxiii. 1946. pp.121. [982.]
lxxxiv. 1947. pp.146. [1114.]
lxxxv. 1948. pp.79. [751.]
lxxxvi. 1949. pp.109. [1028.]
lxxxvii. 1950. pp.93. [903.]
lxxxviii. 1951. pp.86. [729.]
lxxxix. 1952.
xc. 1953. Compiled by A. E. Salisbury. pp.112. [1000.]
xci. 1954. pp.102. [1000.]
xcii. 1955. pp.122. [1100.]
xciii. 1956. Compiled by A. E. Salisbury and Marcia A. Edwards. pp.149. [1200.]
xciv. 1957. pp.144. [1300.]
xcv. 1958. pp.187. [1400.]
xcvi. 1959. pp.96. [1000.]
xcvii. 1960. pp.130. [1500.]

in progress; earlier issues were not published separately; the issues for 1906–1914 also form part of vols.vi–xiv of the International catalogue of scientific literature, N. Zoology.

Animals

ZOOLOGISCHER jahresbericht . . . [1880–1885:
III. abtheilung:] [1880–1882: tunicata, mollusca
(1883–1885: mollusca, brachiopoda; 1886 &c.:
mollusca)].

   1880. Redigirt von J. Vict. Carus. pp.[v].116.
      [600.]
   1881. pp.v.142. [500.]
   1882. Redigirt von Paul Mayer. pp.[iii].148.
      [500.]
   1883. pp.[iii].114. [500.]
   1884. pp.[iii].142. [700.]
   1885. pp.[iii].140. [700.]
   1886. Referent: P. Schiemenz. pp.54. [125.]
   1887. pp.47. [125.]
   1888. pp.66. [150.]
   1889. pp.57. [175.]
   1890. pp.64. [200.]
   1891. pp.61. [175.]
   1892. pp.66. [175.]
   1893. pp.64. [175.]
   1894. pp.70. [150.]
   1895. Referent: Theodor List. pp.51. [125.]
   1896. pp.58. [125.]
   1897. pp.52. [125.]
   1898. pp.46. [125.]
   1899. pp.40. [125.]
   1900. Referent: Karl Hescheler. pp.42. [150.]
   1901. Referent: Walter Stempell. pp.72.

[150.]
1902. pp.62. [150.]
1903. pp.43. [125.]
1904. pp.48. [175.]
1905. pp.34. [150.]
1906. Referent: J. Meisenheimer. pp.32. [150.]
1907. Referent: P. Mayer. pp.22. [125.]
1908. pp.21. [125.]
1909. pp.22. [100.]
1910. pp.30. [150.]
1911. pp.27. [125.]
1912. pp.25. [150.]

*earlier issues and a later issue were not published
separately*.

TH. STUDER, G. ARNSTEIN and A. BROT, Mollus-
ken. Centralkommission für schweizerische landeskunde: Bibliographie der schweizerischen
landeskunde (section IV.6.6.): Bern 1896. pp.xi.12.
[250.]

E. R. SYKES, Digesta malacologica. . . . A summary of the American journal of conchology.
[The Annals and magazine of natural history and
its forerunners. Part I. (1829–1863)]: 1901–1903.
pp.[vii].46+[vii].76. [792.]
no more published.

LIBRARY catalogue (Radley bequest). Malacological society: 1927. pp.23. [500.]

A. MYRA KEEN, An abridged check list and bibliography of west north american marine mollusca. Stanford university: 1937. pp.87. [286.]*

BERNARD C. COTTON and FRANK K. GODFREY, A systematic list of the gastropoda, the marine, freshwater, and land univalve mollusca of South and Central Australia. Malacological society of South Australia: Publication (no.1): Adelaide [1938]. pp.44. [1500.]

BERNARD C. COTTON and FRANK K. GODFREY, A systematic list of the pelecypoda, scaphopoda, cephalopoda, & crepipoda, the bivalve mollusca, tusk-shells, nautilus, squids, octopi, and chitons of South Australia. Malacological society of South Australia: Publication (no.2): Adelaide 1940. pp.32. [750.]

BERNARD C. COTTON and FRANK K. GODFREY, A systematic list of the echinodermata, foraminifera, hydroida, brachiopoda of Southern Australia. Malacological society of Southern Australia: Publication (no.3): Adelaide 1943. pp.36. [750.]

GILBERT RANSON and JACQUELINE PARETIAS, Mollusques perliers et perles (bibliographie). Fondation Albert Ier: Institut océanographique: Bulletin (no.1140): Monaco 1959. pp.43. [800.]

Monotremata.

J. J. FLETCHER, Catalogue of papers and works relating to the mammalian orders, marsupialia and monotremata. [Sydney 1885]. pp.55. [500.]

Mosquitoes.

DON E. EYLES, A critical review of the literature relating to the flight and dispersion habits of anopheline mosquitoes. Public health service: Public health bulletin (no.287): Washington 1944. pp.[ii].39. [150.]

Mule.

G. A. KNYAZEVA, Литература по биологии лошади, осла и мула, по коневодству и конному спорту, изданная в СССР в 1917–1961 гг. (Систематический указатель.) Московская сельскохозяйственная академия имени К. А. Тимирязева: Москва 1962. pp.654. [7500.]

Myriopoda.

WILLIAM FERDINAND ERICHSON, Nomina stystematica generum myriapodum. [Soloduri 1842–1846]. pp.[ii].4. [100.]

forms part of the Nomenclatur zoologicus *of L. Agassiz.*

THE ZOOLOGICAL RECORD. [XII]. Myriopoda [*afterwards:* and protracheata (*afterwards:* prototracheata)].

xiii. 1876. By O. P. Cambridge. single leaf. [12.]

xiv. 1877. pp.2. [20.]

xv. 1878. By E. C. Rye. pp.4. [50.]

xvi. 1879. By W[illiam] F[orsell] Kirby. pp.5. [75.]

xvii. 1880. pp.3. [50.]

xviii. 1881. pp.8. [150.]

xix. 1882. By T. D. Gibson-Carmichael. pp.5. [40.]

xx. 1883. pp.6. [33.]

xxi. 1884. pp.4. [34.]

xxii. 1885. pp.4. [26.]

xxiii. 1886. By R[eginald] Innes Pocock. pp.7. [35.]

xxiv. 1887. pp.13. [50.]

xxv. 1888. pp.6. [25.]

xxvi. 1889. pp.10. [50.]

xxvii. 1890. pp.4. [20.]

xxviii. 1891. pp.6. [15.]

xxix. 1892. pp.7. [20.]

xxx. 1893. pp.11. [50.]

xxxi. [*not published*].

xxxii. 1894–1895. pp.29. [100.]

xxxiii. 1896. By Albert William Brown.

pp.11. [50.]

xxxiv. 1897. pp.19. [70.]

xxxv. 1898. pp.16. [60.]

xxxvi. 1899. pp.17. [42.]

xxxvii. 1900. pp.11. [51.]

xxxviii. 1901. pp.18. [78.]

xxxix. 1902. By W. T. Calman. pp.23. [85.]

xl. 1903. pp.20. [86.]

xli. 1904. By F. Silvestri. pp.17. [75.]

xlii. 1905. pp.16. [53.]

amalgamated with the section on arachnida; the earlier issues were not published separately.

Nerineidae.

W. O. DIETRICH, Gastropoda mesozoica: fam. nerineidae. Fossilium catalogus (1. 31): Berlin 1925. pp.[ii].164. [2500.]

Neuroptera.

L[OUIS AUGUSTE RODOLPHE] AGASSIZ, Nomina systematica generum neuropterorum. [Soloduri 1842–1846]. pp.iv.8.4. [400.]

W. M. SCHØYEN, Fortegnelse over de i Norge hidtil observerede neuroptera planipennia og pseudo-neuroptera. Videnskabs-selskab i Christiania: Forhandlinger (1887, no.13): Christiania 1888. pp.30. [400.]

Noctuidae.

JOHN BERNHARD SMITH, A catalogue, bibliographical and synonymical, of the species of moths of the lepidopterous superfamily, noctuidæ, found in boreal America. National museum: Bulletin (no.44): Washington 1893. pp.424. [7500.]

Odonata.

ERICH SCHMIDT, Bibliographia odonatologica. Ein verzeichnis der titel von schriften über die libellen der erde. [Lieferung 1]. Wien 1933. pp. 114. [1500.]

A–Dobson of the author list only; no more published.

Ornithischia.

OSKAR KUHN, Ornithischia (stegosauriis exclusis). Fossilium catalogus (1. 78): 's-Gravenhage 1936. pp.81. [1000.]

Orthoptera.

L[OUIS JEAN RODOLPHE] AGASSIZ, Nomina systematica generum orthopterorum. [Soloduri 1842–1846]. pp.v.14.2. [500.]

W[ILLIAM] F[ORSELL] KIRBY, A synonymic catalogue of orthoptera. 1904–1910. pp.x.501+viii. 562+ix.674. [30,000.]

Ostracoda.

RAY S[MITH] BASSLER and BETTY KELLETT, Bibliographic index of paleozoic ostracoda. Geological society of America: Special papers (no.1): [Washington] 1934. pp.xiii.500. [7500.]

HENRY V[AN WAGENEN] HOWE, Handbook of ostracod taxonomy. Louisiana state university: Studies: Physical science series (no.1): Baton Rouge [1955]. pp.xviii.389. [2000.]*
— [another edition]. [1962]. pp.xix.366. [2500.]*

Ostrich.

ELIZABETH K. MITCHELL, The ostrich and ostrich farming. A bibliography. University of Cape Town: School of librarianship: Bibliographical series: [Capetown] 1960. pp.vii.55. [313.]*

Otolithi.

O. POSTHUMUS, Otolothi piscium. Fossilium catalogus (1. 24): Berlin 1924. pp.42. [750.]

Oysters.

WILLIAM MARCUS INGRAM and PETER DOUDOROFF, Publications on industrial wastes relating to fish and oysters. A selected bibliography. Public health service: Publication (no.270 = Public health bibliography series, no.10): Washington 1953. pp.ii.28. [114.]★

Pachyodonta.

A. KUTASSY, Pachyodonta mesozoica (rudistis exclusis). Fossilium catalogus (1. 68): 's-Gravenhage 1934. pp.202. [2000.]

Partridge.

CHARLES G. CRISPENS, Quails and partridges of north America. A bibliography. Seattle 1960. pp. xii.125. [2000.]★

Pelmatozoa.

R[AY] S[MITH] BASSLER, Pelmatozoa palaeozoica (generum et genotyporum index et bibliographia). Fossilium catalogus (1. 83): 's-Gravenhage 1938. pp.194. [5000.]

R[AY] S[MITH] BASSLER and MARGARET W. MOODEY, Bibliographic and faunal index of paleozoic pelmatozoan echinoderms. Geological society of America: Special papers (no.45): [New York] 1943. pp.vi.734. [10,000.]

Phyllocarida.

V. VAN STRAELEN and G. SCHMITZ, Crustacea phyllocarida (= archaeostraca). Fossilium catalogus (1. 64): Berlin 1934. pp.246. [3000.]

Phylloxera.

PHYLLOXERA . . . 1950–1959. Science library: Bibliographical series (no.777): [1960]. pp.7. [150.]★

Phytophaga.

J. S. WADE, An annotated bibliography of the Hessian fly, phytophaga destructor (Say). Department of agriculture: Miscellaneous publication (no.198): Washington 1934. pp.100. [1256.]

Pigeon.

A BIBLIOGRAPHY of the books treating on fancy pigeons contained in the library of T. B. Coombe Williams. 1887. pp.20. [150.]
privately printed.

LIST of references on pigeons. Library of Congress: Washington 1916. ff.2. [20.]★

BRIEF list upon carrier and homing pigeons. Library of Congress: Washington 1919. ff.2. [35.]★

WERNER K. G. MÖBES, Bibliographie der tauben. Halle 1945. pp.3–191. [1500.]

Pisces. *see* **Fish.**

Placodontia.

OSKAR KUHN, Placodontia. Fossilium catalogus (1. 62): Berlin 1933. pp.15. [100.]

Animals

Poikilotherms.

ANNE RAFFY, Bibliographie relative à la mesure du métabolisme respiratoire des poïkilothermes aquatiques. Institut océanographique: Bulletin (no.623): Monaco 1933. pp.20. [300.]

Polypi.

L[OUIS AUGUSTE RODOLPHE] AGASSIZ, Nomina systematica generum polyporum. [Soloduri 1842–1846]. pp.viii.28.4. [1100.]

Porifera.

S[AMUEL] J[OSEPH] MACKIE, An illustrated catalogue of british fossil sponges. [Part II. Bibliography]. 1866. pp.32. [40.]
no more published.

THE ZOOLOGICAL RECORD. [XVII; *afterwards:* III]. Spongozoa [*afterwards:* Spongida; Spongiida; Spongiæ; Porifera or Spongida; Porifera].

   xiii. 1876. By C[hristian] F[rederik] Lutken. pp.6. [50.]
   xiv. 1877. By Stuart O. Ridley. pp.10. [50.]
   xv. 1878. pp.15. [50.]
   xvi. 1879. pp.16. [50.]
   xvii. 1880. pp.23. [50.]
   xviii. 1881. pp.14. [50.]

xix. 1882. pp.23. [50.]

xx. 1883. By W[illiam] J[ohnson] Sollas. pp.11. [50.]

xxi. 1884. pp.11. [53.]

xxii. 1885. pp.19. [52.]

xxiii. 1886. pp.27. [43.]

xxiv. 1887. pp.19. [50.]

xxv. 1888. By Oswald H. Latter. pp.10. [33.]

xxvi. 1889. pp.20. [48.]

xxvii. 1890. By E. A. Minchin. pp.19. [50.]

xxviii. 1891. pp.33. [68.]

xxix. 1892. By R[ichard] Hanitsch. pp.24. [82.]

xxx. 1893. pp.19. [72.]

xxxi. 1894. pp.12. [53.]

xxxii. 1895. By R[obert] von Lendenfeld. pp.18. [70.]

xxxiii. 1896. pp.15. [57.]

xxxiv. 1897. pp.16. [42.]

xxxv. 1898. pp.19. [77.]

xxxvi. 1899. pp.12. [48.]

xxxvii. 1900. By E. A. Minchin. pp.55. [83.]

xxxviii. 1901. pp.43. [69.]

xxxix. 1902. pp.38. [76.]

xl. 1903. pp.32. [59.]

xli. 1904. By W. Woodland. pp.26. [60.]

xlii. 1905. By E. A. Minchin. pp.26. [48.]

xliii. 1906. By I. B. J. Sollas. pp.7. [60.]

xliv. 1907. pp.10. [44.]

xlv. 1908. pp.9. [54.]

xlvi. 1909. pp.8. [54.]

xlvii. 1910. pp.10. [45.]

xlviii. 1911. pp.8. [45.]

xlix. 1912. pp.10. [46.]

l. 1913. By R. W. Harold Row. pp.8. [50.]

li. 1914. pp.8. [36.]

lii. 1915. pp.3. [46.]

liii. 1916. pp.7. [31.]

liv. 1917. By H. M. Woodcock. pp.4. [18.]

lv. 1918. pp.4. [16.]

lvi. 1919. pp.4. [22.]

lvii. 1920. pp.6. [30.]

lviii. 1921. pp.4. [16.]

lix. 1922. pp.4. [26.]

lx. 1923. pp.3. [19.]

lxi. 1924. pp.7. [18.]

lxii. 1925. By M[aurice] Burton. pp.8. [59.]

lxiii. 1926. pp.6. [54.]

lxiv. 1927. pp.11. [46.]

lxv. 1928. pp.6. [55.]

lxvi. 1929. pp.12. [65.]

lxvii. 1930. pp.7. [65.]

lxviii. 1931. pp.6. [50.]

lxix. 1932. pp.12. [65.]

lxx. 1933. pp.8. [65.]
lxxi. 1934. pp.8. [64.]
lxxii. 1935. pp.7. [55.]
lxxiii. 1936. pp.14. [58.]
lxxiv. 1937. pp.7. [66.]
lxxv. 1938. pp.6. [43.]
lxxvi. 1939. pp.5. [57.]
lxxvii. 1940. pp.3. [22.]
lxxviii. 1941. pp.5. [28.]
lxxix. 1942. pp.3. [15.]
lxxx. 1943. pp.2. [17.]
lxxxi. 1944. pp.4. [36.]
lxxxii. 1945. pp.7. [77.]
lxxxiii. 1946. pp.4. [45.]
lxxxiv. 1947. pp.2. [15.]
lxxxv. 1948. pp.4. [39.]
lxxxvi. 1949. pp.4. [48.]
lxxxvii. 1950. pp.7. [66.]
lxxxviii. 1951. pp.4. [41.]
lxxxix. 1952.
xc. 1953. pp.4. [32.]
xci. 1954. pp.2. [16.]
xcii. 1955. pp.2 [14.]
xciii. 1956. pp.6. [72.]
xciv. 1957. pp.10. [60.]
xcv. 1958. pp.8. [60.]
xcvi. 1959. pp.4. [30.]

xcvii. 1960. Compiled by S. Ware. pp.14. [150.]

in progress; the earlier issues were not published separately; the issues for 1906–1914 also form vols.vi–xiv of the International catalogue of scientific literature, N. Zoology.

D'ARCY W[ENTWORTH] THOMPSON, A bibliography of protozoa, sponges, cœlenterata, and worms, including also the polyzoa, brachiopoda and tunicata, for the years 1861–1883. Cambridge 1885. pp.viii.284. [6000.]

ZOOLOGISCHER Jahresbericht.... Porifera. Zoologische station zu Neapel: Berlin.

1886. Referent: G. C. J. Vosmaer. pp.5. [15.]
1887. pp.9. [25.]
1888. pp.9. [20.]
1889. pp.8. [50.]
1890. pp.6. [35.]
1891. pp.8. [35.]
1892. pp.12. [50.]
1893. Referent: G. P. Bidder. pp.14. [35.]
1894. pp.12. [40.]
1895. Referent: B. Nöldeke. pp.5. [30.]
1896. pp.7. [25.]
1897. pp.6. [25.]
1898. pp.9. [50.]

1899. Referent: O[tto] Maas. pp.8. [25.]

1900. pp.7. [25.]

1901. pp.6. [25.]

1902. pp.6. [25.]

1903. pp.6. [25.]

1904. pp.9. [20.]

1905. pp.10. [25.]

1906. pp.5. [25.]

1907. pp.10. [25.]

1908. pp.9. [30.]

1909. pp.9. [30.]

1910. Referent: E. Hentschel. pp.7. [25.]

1911. pp.7. [30.]

1912. pp.6. [35.]

earlier issues and a later issue were not published separately.

G[UALTHERUS] C[AREL] J[ACOB] VOSMAER, Bibliography of sponges, 1551–1913. . . . Edited by G[eorge] P[arker] Bidder and C[atalina] S[usanne] Vosmaer-Röell. Cambridge 1928. pp.xii.234. [3750.]

Porpoise.

WERNER K. G. MOEBES, Bibliographie des kaninchens. Nebst anhang I. Das frettchen, II. Das meerschweinchen. Halle &c. [1946]. pp.xvi.318. [porpoise: 50.]

Animals

Primates. *see* **Zoology.**

Prochordata. *see* **Tunicata.**

Protochordata. *see* **Tunicata.**

Protorosauria.

OSKAR KUHN, Protorosauria, mesosauria. Fossilium catalogus (I. 85): 's-Gravenhage 1939. pp. [ii].8.6. [protorosauria: 75.]

Prototracheata. *see* **Myriopoda.**

Protozoa. [*see also* **Bacteriology.**]

THE ZOOLOGICAL RECORD. [XVIII; *afterwards:* II.] Protozoa.

> xiii. 1876. By C[hristian[F[rederik] Lütken. pp.12. [100.]
>
> xiv. 1877. By Stuart O. Ridley. pp.13. [100.]
>
> xv. 1878. pp.18. [150.]
>
> xvi. 1879. pp.20. [150.]
>
> xvii. 1880. pp.22. [200.]
>
> xviii. 1881. pp.35. [300.]
>
> xix. 1882. pp.16. [150.]
>
> xx. 1883. By Alfred C. Haddon. pp.13. [150.]
>
> xxi. 1884. pp.17. [125.]
>
> xxii. 1885. pp.15. [105.]
>
> xxiii. 1886. By J. Arthur Thomson. pp.21. [100.]
>
> xxiv. 1887. pp.22. [137.]
>
> xxv. 1888. By Cecil Warburton. pp.22. [150.]

xxvi. 1889. pp.16. [100.]

xxvii. 1890. pp.14. [100.]

xxviii. 1891. pp.13. [100.]

xxix. 1892. By R[ichard] Hanitsch. pp.32. [165.]

xxx. 1893. pp.34. [266.]

xxxi. 1894. pp.22. [279.]

xxxii. 1895. By W. Fraser Hume and Frederick Chapman. pp.31. [229.]

xxxiii. 1896. By Albert William Brown. pp. 24. [200.]

xxxiv. 1897. pp.22. [171.]

xxxv. 1898. pp.22. [208.]

xxxvi. 1899. pp.26. [148.]

xxxvii. 1900. pp.28. [167.]

xxxviii. 1901. pp.36. [288.]

xxxix. 1902. By H. M. Woodcock. pp.88. [441.]

xl. 1903. pp.72. [432.]

xli. 1904. pp.60. [371.]

xlii. 1905. pp.76. [522.]

xliii. 1906. pp.48. [400.]

xliv. 1907. pp.59. [476.]

xlv. 1908. pp.60. [492.]

xlvi. 1909. pp.64. [554.]

xlvii. 1910. pp.72. [559.]

xlviii. 1911. pp.67. [523.]

Animals

lxxv. 1938. pp.85. [881.]

lxxvi. 1939. By C. A. Hoare. pp.108. [1119.]

lxxvii. 1940. pp.70. [697.]

lxxviii. 1941. pp.49. [512.]

lxxix. 1942. pp.69. [723.]

lxxx. 1943. pp.60. [706.]

lxxxi. 1944. pp.63. [747.]

lxxxii. 1945. pp.66. [745.]

lxxxiii. 1946. pp.114. [1176.]

lxxxiv. 1947. pp.85. [889.]

lxxxv. 1948. pp.91. [1038.]

lxxxvi. 1949. pp.110. [1127.]

lxxxvii. 1950. pp.127. [1268.]

lxxxviii. 1951. pp.105. [1126.]

lxxxix. 1952.

xc. 1953. Compiled by Cecil A. Hoare and R. H. Cummings. pp.127. [1445.]

xci. 1954. pp.123. [1408.]

xcii. 1955. pp.119. [1300.]

xciii. 1956. Compiled by R. H. Cummings and R. A. Neal. pp.124. [1300.]

xciv. 1957. Compiled by R. A. Neal. pp.94. [1200.]

xcv. 1958. Compiled by R. A. Neal and R. H. Cummings. pp.148. [1600.]

xcvi. 1959. Compiled by R. A. Neal and R. S. J. Hawes. pp.72. [1200.]

xcvii. 1960. Compiled by R. A. Neal [*and others*]. pp.129. [2500.]

in progress; earlier issues were not published separately; the issues for 1906–1914 also form part of vols.vi–xiv of the International catalogue of scientific literature, N. Zoology.

D'ARCY W. THOMPSON, A bibliography of protozoa, sponges, cœlenterata, and worms, including also the polyzoa, brachiopoda and tunicata, for the years 1861–1883. Cambridge 1885. pp.viii.284. [6000.]

ZOOLOGISCHER jahresbericht. . . . Protozoa. Zoologische station zu Neapel: Berlin.

 1886. Referent: Karl Brandt. pp.12. [60.]
 1887. Referent: J. van Rees. pp.18. [75.]
 1888. pp.24. [125.]
 1889. pp.27. [100.]
 1890. pp.25. [100.]
 1891. Referent: P. Schiemenz. pp.27. [150.]
 1892. pp.35. [200.]
 1893. pp.31. [200.]
 1894. Referent: Theodor List. pp.29. [125.]
 1895. pp.44. [125.]
 1896. pp.31. [125.]
 1897. pp.34. [125.]
 1898. pp.29. [125.]

1899. pp.34. [175.]
1900. Referent: Paul Mayer. pp.18. [50.]
1901. pp.16. [150.]
1902. pp.29. [200.]
1903. pp.26. [200.]
1904. pp.38. [275.]
1905. pp.29. [275.]
1906. pp.31. [300.]
1907. Referent: J. Gross. pp.53. [300.]
1908. pp.52. [400.]
1909. pp.74. [500.]
1910. pp.66. [450.]
1911. pp.61. [600.]
1912. pp.54. [500.]

earlier issues and a later issue were not published separately.

Protracheata. *see* **Myriopoda.**

Psyllida.

G. AULMANN, Psyllidarum catalogus. Berlin 1913. pp.92. [2000.]

Pterosauria.

F. PLIENINGER, Pterosauria. Fossilium catalogus (I. 45): Berlin 1929. pp.84. [1250.]

Pulex.

HUGO HAYN and ALFRED N. GOTENDORF, Floh-litteratur (de pulicibus) des in- und auslandes, vom

XVI. jahrhundert bis zur neuzeit. [*s.l.*] 1913. pp. [ii].36. [200.]

a copy in the Bodleian library contains numerous ms. notes by John Hodgkin.

Quails.

CHARLES G. CRISPENS, Quails and partridges of north America. A bibliography. Seattle 1960. pp.xii.125. [2000.]★

Rabbits.

LIST of references on rat an rabbit depredations and their control. Library of Congress: [Washington] 1921. ff.7. [60.]★

BIBLIOGRAPHY on the genetics and sex physiology of the rabbit. Imperial bureau of animal genetics: University of Edinburgh: Edinburgh 1931. ff.[i].34. [300.]★

REFERENCES to the inheritance of coat-colour in rabbits. Science library: Bibliographical series (no.51): [1932]. ff.[i].3. [32.]★

CONTROL and extermination of moles and rabbits. Science library: Bibliographical series (no.331): 1937. ff.6. [136.]★

CARLTON M[ARTIN] HERMAN, The rabbit used in disease research. A selected bibliography, including the spontaneous diseases of rabbits. Fish and wildlife service: Chicago 1942. pp.v.519. [4859.]★

WERNER K. G. MOEBES, Bibliographie des kaninchens. Nebst anhang I. Das frettchen, II. Das meerschweinchen. Halle &c. [1946]. pp.xvi.318. [1500.]

LAURA I[SABEL] MAKEPEACE, Rabbits. A subject bibliography. Bibliographical center for research: Rocky mountain region: Special bibliography (no.3): Denver 1956. pp.xii.81. [3485.]★

Rats.

LIST of references on rat and rabbit depredations and their control. Library of Congress: [Washington] 1921. ff.7. [60.]★

Reindeer.

LIST of references on the reindeer industry. Library of Congress: Washington 1916. ff.4. [47.]★

Reptilia. [*see also* Fish.]

L[OUIS CHARLES RODOLPHE] AGASSIZ, Nomina systematica generum reptilium. [Soloduri 1842–1846]. pp.x.48.8. [2000.]

[S. JOURDAIN], Travaux d'erpétologie et d'ichthyologie de S. Jourdain. Montpellier 1872. pp.10. [17.]

THE ZOOLOGICAL RECORD. [IV; *afterwards:* XVI;

xiv; xv; xvi]. Reptilia [*afterwards:* and batrachia; Amphibia and reptilia].

    xiii. 1876. By A. W. E. O'Shaughnessy. pp.18. [300.]

    xiv. 1877. pp.14. [250.]

    xv. 1878. pp.15. [250.]

    xvi. 1879. pp.19. [300.]

    xvii. 1880. By G. A. Boulenger. pp.13. [250.]

    xviii. 1881. pp.16. [300.]

    xix. 1882. pp.28. [500.]

    xx. 1883. pp.24. [400.]

    xxi. 1884. pp.19. [300.]

    xxii. 1885. pp.26. [400.]

    xxiii. 1886. pp.24. [300.]

    xxiv. 1887. pp.34. [300.]

    xxv. 1888. pp.28. [250.]

    xxvi. 1889. pp.23. [250.]

    xxvii. 1890. pp.28. [250.]

    xxviii. 1891. pp.24. [200.]

    xxix. 1892. pp.41. [300.]

    xxx. 1893. pp.38. [300.]

    xxxi. 1894. pp.44. [300.]

    xxxii. 1895. pp.35. [200.]

    xxxiii. 1896. pp.38. [300.]

    xxxiv. 1897. pp.33. [250.]

    xxxv. 1898. pp.29. [250.]

    xxxvi. 1899. pp.31. [250.]

xxxvii. 1900. pp.31. [250.]

xxxviii. 1901. pp.35. [250.]

xxxix. 1902. pp.32. [250.]

xl. 1903. pp.38. [300.]

xli. 1904. pp.40. [350.]

xlii. 1905. By I. B. J. Sollas. pp.40. [429.]

xliii. 1906. pp.39. [350.]

xliv. 1907. pp.35. [403.]

xlv. 1908. pp.34. [373.]

xlvi. 1909. By C. L. Boulenger. pp.35. [354.]

xlvii. 1910. pp.34. [376.]

xlviii. 1911. pp.34. [382.]

xlix. 1912. pp.32. [351.]

l. 1913. By [Charles] Tate Regan. pp.31. [335.]

li. 1914. pp.20. [255.]

lii. 1915. pp.19. [238.]

liii. 1916. pp.14. [172.]

liv. 1917. pp.13. [138.]

lv. 1918. pp.12. [146.]

lvi. 1919. pp.12. [131.]

lvii. 1920. By Joan B. Procter. pp.23. [274.]

lviii. 1921. pp.28. [355.]

lix. 1922. By W. L. Sclater. pp.32. [406.]

lx. 1923. By J. B. Procter. pp.41. [530.]

lxi. 1924. pp.41. [558.]

lxii. 1925. By Stanley S. Flower. pp.61. [779.]

lxxxvii. 1950. pp.115. [880.]

lxxxviii. 1951. pp.138. [1233.]

lxxxix. 1952.

xc. 1953. pp.88. [500.]

xci. 1954. pp.85. [500.]

xcii. 1955. pp.101. [500.]

xciii. 1956. pp.122. [700.]

xciv. 1957. pp.94. [600.]

xcv. 1958. pp.93. [600.]

xcvi. 1959. pp.56. [700.]

xcvii. 1960. Compiled by J. C. Battersby, Marcia A. Edwards and Pauline Curds. pp.70. [650.]

xcviii. 1961. Compiled by M. A. Edwards and P. Curds. pp.79. [750.]

in progress; earlier issues were not published separately; the issues for 1906–1914 also form part of vols.vi–xiv of the International catalogue of scientific literature, N. Zoology.

H. FISCHER-SIGWART, Reptilien und amphibien. Centralkommission für schweizerische landeskunde: Bibliographie der schweizerischen landeskunde (section IV.6.5γ): Bern 1898. pp.xii.27. [500.]

BARON F. VON NOPCSA [VON FELSÖ-SZILVÁS], Osteologia reptilium fossilium et recentium.

Fossilium catalogus (I. 27): Berlin 1926. pp.[ii]. 391. [7500.]

—— Appendix. [Second supplement]. . . . (I. 50): 1931. pp.62. [900.]
the first supplement forms part of the main work.

[HOWARD H. PECKHAM, HELEN T. GAIGE and CARL L. HUBBS], Ichthyologia et herpetologia americana. A guide to an exhibition in the William L. Clements library illustrating the development of knowledge of american fishes, amphibians and reptiles. William L. Clements library: Bulletin (no.xxv): Ann Arbor 1936. pp. [ii].22. [46.]

HOBART M[UIR] SMITH and EDWARD H[ARRISON] TAYLOR, An annotated checklist and key to the reptiles of Mexico exclusive of the snakes. United States national museum: Bulletin (no.199): Washington 1950. pp.v.253. [250.]

Rhynchocephalia.

OSKAR KUHN, Rhynchocephalia (eosuchia). Fossilium catalogus (I. 71): 's-Gravenhage 1935. pp.39. [500.]

Rodentia.

[SIR] J[OHN] R[EEVES] ELLERMAN, The families and genera of living rodents. . . . With a list of

named forms (1558–1936) by R. W. Hayman and
G. W. C. Holt. British museum (natural history).

    i. Rodents other than muridae. 1940. pp.
      xxvi.689. [4000.]

    ii. Family muridae. 1941. pp.xii.690. [4000.]

Rotatoria.

L. AGASSIZ, Nomina systematica generum rota-
torium. [Soloduri 1842–1846]. pp.[iii].5. [150.]

Roundworms.

CH[ARLES] WARDELL STILES and ALBERT HASSALL,
Index-catalogue of medical and veterinary zoology. Subjects: roundworms (nematoda, gor-
diacea, and acanthocephali) and the diseases they
cause. Hygienic laboratory: Bulletin (no.114):
Washington 1920. pp.886. [75,000.]

—— [another edition]. By Mildred A[nn]
Doss. 1963 &c.

    in progress.

Rudistae.

O. KÜHN, Rudistae. Fossilium catalogus (I. 54):
Berlin 1932. pp.200. [3000.]

Saurischia.

F. DE HUENE, Saurischia et ornithischia triadica.
('Dinosauria' triadica). Fossilium catalogus (I. 4):
Berlin 1914. pp.21. [150.]

OSKAR KUHN, Saurischia. Fossilium catalogus (1. 87): 's-Gravenhage 1939. pp.124. [1500.]

Sauropterygia.

OSKAR KUHN, Sauropterygia. Fossilium catalogus (1. 69): 's-Gravenhage 1934. pp.127. [1750.]

Scaphopoda.

G. HABER, Gastropoda, amphineura et scaphopoda jurassica. Fossilium catalogus (1. 53, 65 &c.): Berlin 1932 &c.

Scorpion.

THE REPRODUCTIVE organs and development of scorpions. Science library: Bibliographical series (no.16): [1930]. ff.[i].3. [34.]★
— 2nd edition . . . (no.109): 1934. ff.[7]. [34.]★

Seal.

RECENT references on fur seals. Library of Congress: Washington 1914. ff.3. [32.]★

RUDOLF FRITZSCHE, Robben. Reichs-zentrale für pelztier- und rauchwaren-forschung: Literatursammlung für tierkunde und tierzucht: Leipzig 1939. pp.16. [300.]

Siphonaptera.

W[ILLIA]M L[IVINGSTON] JELLISON and NEWELL E[MANUEL] GOOD, Index to the literature of siphonaptera of north America. Public health service: National institute of health bulletin (no. 178): Washington 1942. pp.iv.193. [1500.]

Snakes.

LIST of references on the serpent in folklore and mythology. Library of Congress: Washington 1916. ff.6. [68.]*

SNAKES. Science library: Bibliographical series (no.63): [1932]. ff.[i].4. [57.]*

THE AFFINITIES and genealogical tree of the snake families. . . . The scalation of various snakes. Second edition. Science library: Bibliographical series (no.109): 1934. ff.[7]. [40.]*

EMBRYOLOGY of snakes. Science library: Bibliographical series (no.149): 1934. ff.[i].2. [35.]*

HOBART M[UIR] SMITH and EDWARD H[ARRISON] TAYLOR, An annotated checklist and key to the snakes of Mexico. Smithsonian institution: United States national museum: Bulletin (no.187): Washington 1945. pp.iv.239. [1500.]

Spiders. *see* **Arachnida.**
Sponges. *see* **Porifera.**
Sporozoa.

[PAUL OCTAVE] HAGENMÜLLER, Bibliotheca

Animals

sporozoologica. Annales du Musée d'histoire naturelle de Marseille: Bulletin (2nd ser., vol.ii, supplement): Marseille 1899. pp.233. [2000.]

Squamata.

OSKAR KUHN, Squamata: lacertilia et ophidia. Fossilium catalogus (1. 86): 's-Gravenhage 1939. pp.[ii].89.33. [1500.]

Stegosauria.

E. HENNIG, Stegosauria. Fossilium catalogus (1. 9): Berlin 1915. pp.16. [87.]

Stelleroidea.

CHARLES SCHUCHERT, Stelleroidea palaeozoica. Fossilium catalogus (1. 3): Berlin 1914. pp.53. [1000.]

Strepsiptera.

WILLIAM FERDINAND ERICHSON, Nomina systematica generum strepsipterorum. [Soloduri 1842–1846]. pp.[3]. [8.]
forms part of the Nomenclator zoologicus *of Louis Agassiz.*

Suctoria.

WILLIAM FERDINAND ERICHSON, Nomina systematica generum suctoriorum. [Soloduri 1842–1846]. pp.[3]. [12.]

forms part of the Nomenclator zoologicus *of Louis Agassiz.*

Termites.

THOMAS E[LLIOTT] SNYDER, Annotated subject-heading bibliography of termites 1350 B.C. to A.D. 1954. Smithsonian institution: Smithsonian miscellaneous collections (vol.130): Washington 1956. pp.iii.305. [5500.]

Thecodontia.

OSKAR KUHN, Theodontia. Fossilium catalogus (1. 58): Berlin 1933. pp.32. [500.]
— — Supplementum . . . (1. 63): 1934. pp.iv. [50.]

Theromorpha.

OSKAR KUHN, Cotylosauria et theromorpha. Fossilium catalogus (1. 79): 's-Gravenhage 1937. pp.209. [3000.]

Thysanoptera.

WILLIAM FERDINAND ERICHSON, Nomina systematica generum thysanopterorum. [Soloduri 1842–1846]. pp.[3]. [19.]
forms part of the Nomenclator zoologicus *of Louis Agassiz.*

Thysanura.

WILLIAM FERDINAND ERICHSON, Nomina systematica generum thysanurorum. [Soloduri 1842–1846]. pp.[ii].2. [27.]
forms part of the Nomenclator zoologicus *of Louis Agassiz.*

Trematoda.

CH[ARLES] WARDELL STILES and ALBERT HASSALL, Index-catalogue of medical and veterinary zoology. Subjects: trematoda and trematode diseases. Hygienic laboratory: Bulletin (no.37): Washington 1908. pp.401. [25,000.]

Trigoniidae.

W. DEECKE, Trigoniidae mesozoicae (myophoria exclusa). Fossilium catalogus (I. 30): Berlin 1925. pp.[ii].306. [5000.]

Trilobitae.

RUDOLF and E. RICHTER, Trilobitae neodevonici. Fossilium catalogus (I. 37): Berlin 1928. pp.[ii]. 160. [2000.]

THE ZOOLOGICAL record. XI. Trilobita.
    lxxv. 1938. By C. J. Stubblefield. pp.32. [97.]
    lxxvi. 1939. pp.21. [74.]
    lxxvii. 1940. pp.16. [46.]
    lxxviii. 1941. pp.14. [35.]

lxxix. 1942. pp.18. [38.]
lxxx. 1943. pp.12. [49.]
lxxxi. 1944. pp.20. [77.]
lxxxii. 1945. pp.27. [130.]
lxxxiii. 1946. pp.23. [85.]
lxxxiv. 1947. pp.26. [78.]
lxxxv. 1948. pp.31. [88.]
lxxxvi. 1949. pp.18. [55.]
lxxxvii. 1950. pp.31. [84.]
lxxxviii. 1951. pp.37. [78.]
lxxxix. 1952. [*none.*]
xc. 1953. Compiled by J. T. Temple. pp.48. [121.]
xci. 1954. pp.90. [84.]
xcii. 1955. pp.59. [80.]
xciii. 1956. pp.69. [150.]
xciv. 1957. pp.68. [150.]
xcv. 1958. pp.86. [275.]
xcvi. 1959. pp.32. [200.]
xcvii. 1960. pp.43. [150.]
xcviii. 1961. pp.35. [200.]

in progress; the earlier volumes contained no separate section devoted to trilobitae, which were included in that dealing with Arachnida.

Trionychia.

K. HUMMEL, Trionychia fossilia. Fossilium catalogus (I. 52): Berlin 1932. pp.106. [2000.]

Animals

Trochus.

R[ENÉ] GAIL and L. DEVAMBEZ, Bibliographie analytique du Troca (Trochus niloticus, Linn.). Commission du Pacifique sud: Nouméa 1958. pp.[i].iii.20. [50.]

Tuna.

GENEVIEVE CORWIN, A bibliography of the tunas. California state fisheries laboratory: Contribution (no.87): [Sacramento 1930]. pp.104. [900.]

BELL M. SHIMADA, An annotated bibliography on the biology of Pacific tunas. Fish and wildlife service: Fishery bulletin (no.58): Washington 1951. pp.58. [500.]

Tunicata.

ZOOLOGISCHER jahresbericht. . . . III. Abtheilung: tunicata, mollusca. Zoologische station zu Neapel: Leipzig.

> 1880. Redigirt von J. Vict. Carus. pp.[v].116. [600.]
>
> 1881. pp.v.142. [500.]
>
> 1882. Redigirt von Paul Mayer. pp.[iii].148. [500.]
>
> [*continued as:*]

Zoologischer jahresbericht. . . . IV. Abtheilung: tunicata, vertebrata.

1883. pp.iv.334. [1500.]
1884. pp.iv.414. [1250.]
1885. pp.iv.336. [1500.]
[*continued as:*]
Zoologischer jahresbericht. . . . Tunicata.
1886. Referent: A. della Valle. pp.8. [25.]
1887. pp.4. [25.]
1888. pp.6. [15.]
1889. pp.6. [10.]
1890. pp.7. [10.]
1891. pp.13. [25.]
1892. pp.7. [25.]
1893. pp.34. [50.]
1894. pp.13. [25.]
1895. pp.14. [25.]
1896. pp.12. [25.]
1897. pp.5. [15.]
1898. pp.6. [25.]
1899. pp.10. [30.]
1900. pp.9. [30.]
1901. pp.5. [20.]
1902. pp.6. [20.]
1903. pp.8. [20.]
1904. pp.13. [30.]
1905. pp.12. [40.]
1906. pp.5. [25.]
1907. pp.5. [25.]

1908. pp.8. [25.]
1909. pp.5. [25.]
1910. pp.5. [25.]
1911. pp.5. [25.]
1912. pp.7. [45.]

earlier issues and a later issue were not published separately.

THE ZOOLOGICAL RECORD. [VI; *afterwards:* XII; XIII; XIV]. Tunicata [*afterwards:* Prochordata; Protochordata].

xxi. 1884. By Eduard von Martens. pp.6. [25.]

xxii. 1885. By W. A. Herdman. pp.8. [30.]

xxiii. 1886. pp.8. [21.]

xxiv. 1887. pp.5. [21.]

xxv. 1888. pp.5. [11.]

xxvi. 1889. pp.5. [13.]

xxvii. 1890. pp.6. [15.]

xxviii. 1891. pp.6. [21.]

xxix. 1892. pp.7. [31.]

xxx. 1893. pp.8. [34.]

xxxi. 1894. pp.8. [40.]

xxxii. 1895. pp.7. [20.]

xxxiii. 1896. pp.8. [20.]

xxxiv. 1897. pp.4. [15.]

xxxv. 1898. pp.8. [30.]

xxxvi. 1899. pp.7. [29.]

Animals

xxxvii. 1900. pp.6. [23.]

xxxviii. 1901. By Alice L. Embleton. pp.5. [31.]

xxxix. 1902. By W[illiam] T[homas] Calman. pp.8. [38.]

xl. 1903. pp.7. [31.]

xli. 1904. pp.7. [29.]

xlii. 1905. pp.6. [42.]

xliii. 1906. By I. B. J. Sollas. pp.7. [40.]

xliv. 1907. pp.7. [47.]

xlv. 1908. pp.8. [53.]

xlvi. 1909. pp.8. [45.]

xlvii. 1910. pp.6. [26.]

xlviii. 1911. pp.4. [33.]

xlix. 1912. pp.7. [48.]

l. 1913. By C[harles] Tate Regan. pp.7. [43.]

li. 1914. pp.4. [30.]

lii. 1915. pp.3. [18.]

liii. 1916. pp.3. [14.]

liv. 1917. pp.2. [5.]

lv. 1918. pp.3. [8.]

lvi. 1919. pp.3. [8].

lvii. 1920. pp.3. [15.]

lviii. 1921. By J. R. Norman. pp.3. [18.]

lix. 1922. pp.4. [28.]

lx. 1923. pp.4. [35.]

lxi. 1924. pp.4. [45.]

lxii. 1925. pp.3. [18.]

Animals

lxiii. 1926. pp.6. [55.]
lxiv. 1927. pp.6. [47.]
lxv. 1928. By M. Burton. pp.6. [58.]
lxvi. 1929. pp.4. [46.]
lxvii. 1930. pp.5. [42.]
lxviii. 1931. pp.5. [50.]
lxix. 1932. pp.5. [7].
lxx. 1933. pp.4. [49.]
lxxi. 1934. pp.6. [55.]
lxxii. 1935. pp.5. [49.]
lxxiii. 1936. pp.5. [48.]
lxxiv. 1937. pp.6. [69.]
lxxv. 1938. pp.4. [42.]
lxxvi. 1939. pp.3. [42.]
lxxvii. 1940. pp.2. [12.]
lxxviii. 1941. pp.3. [23.]
lxxix. 1942. [*none.*]
lxxx. 1943. pp.2. [15.]
lxxxi. 1944. pp.2. [23.]
lxxxii. 1945. pp.6. [71.]
lxxxiii. 1946. pp.2. [18.]
lxxxiv. 1947. pp.2. [13.]
lxxxv. 1948. pp.4. [50.]
lxxxvi. 1949. pp.4. [46.]
lxxxvii. 1950. pp.6. [78.]
lxxxviii. 1951. pp.4. [51.]
lxxxix. 1952. pp.6. [63.]

xc. 1953. Compiled by M[aurice] Burton. pp.4. [35.]

xci. 1954. Compiled by D. B. Carlisle. pp.9. [74.]

xcii. 1955. pp.10. [90.]

xciii. 1956. pp.14. [30.]

xciv. 1957. pp.13. [100.]

xcv. 1958. pp.15. [100.]

xcvi. 1959. pp.13. [150.]

xcvii. 1960. pp.21. [175.]

xcviii. 1961. pp.22. [150.]

in progress; earlier issues were not published separately; the issues for 1906–1914 also form part of vols.vi–xiv of the International catalogue of scientific literature, N. Zoology.

JOHN HOPKINSON, A bibliography of the tunicata, 1469–1910. Ray society: 1913. pp.xii.288. [4000.]

Turtle.

BRIEF list of references on turtles. Library of Congress: Washington 1922. ff.7. [71.]★

Tyroglyphus.

THE FLOUR mite, Tyroglyphus (aleurobius) farinae. Science library: Bibliographical series (no.170): 1935. ff.[i].2. [25.]★

Vermes.

L[OUIS JEAN RODOLPHE] AGASSIZ, Nomina systematica generum vermium, (annulatorum, turbellarium et entozoorum). [Soloduri 1842–1846]. pp.viii.14.5. [650.]

THE ZOOLOGICAL RECORD. [xv; *afterwards:* VI]. Vermes [*afterwards:* Vermidea; Vermes].

    xiii. 1876. By C[hristian] F[rederik] Lütken. pp.22. [400.]

    xiv. 1877. By F. Jeffrey Bell. pp.21. [400.]

    xv. 1878. pp.17. [300.]

    xvi. 1879. pp.18. [300.]

    xvii. 1880. pp.15. [300.]

    xviii. 1881. pp.13. [300.]

    xix. 1882. pp.16. [300.]

    xx. 1883. pp.14. [300.]

    xxi. 1884. pp.17. [100.]

    xxii. 1885. pp.22. [117.]

    xxiii. 1886. pp.16. [134.]

    xxiv. 1887. By G. Herbert Fowler. pp.34. [250.]

    xxv. 1888. By [sir] P[eter] Chalmers Mitchell. pp.31. [250.]

    xxvi. 1889. pp.35. [250.]

    xxvii. 1890. pp.27. [200.]

    xxviii. 1891. By Arthur Willey. pp.50. [250.]

    xxix. 1892. By Florence Buchanan. pp.88.

Animals

[500.]

xxx. 1893. pp.63. [269.]

xxxi. 1894. pp.45. [253.]

xxxii. 1895. pp.47. [271.]

xxxiii. 1896. pp.49. [251.]

xxxiv. 1897. pp.50. [267.]

xxxv. 1898. pp.60. [333.]

xxxvi. 1899. By Arthur Willey. pp.50. [418.]

xxxvii. 1900. pp.50. [345.]

xxxviii. 1901. By Alice L. Embleton. pp.63. [412.]

xxxix. 1902. pp.72. [486.]

xl. 1903. pp.60. [477.]

xli. 1904. pp.57. [484.]

xlii. 1905. By Cora B. Sanders. pp.73. [456.]

xliii. 1906. By F. A. Potts. pp.51. [400.]

xliv. 1907. pp.48. [426.]

xlv. 1908. pp.47. [442.]

xlvi. 1909. By H. M. Woodcock. pp.56. [498.]

xlvii. 1910. pp.56. [476.]

xlviii. 1911. pp.52. [400.]

xlix. 1912. By J. S. Dunkerly. pp.56. [534.]

l. 1913. pp.51. [471.]

li. 1914. By H. M. Woodcock. pp.37. [364.]

lii. 1915. By R. W. Harold Row. pp.40. [391.]

liii. 1916. pp.31. [269.]

Animals

lxxxi. 1944. By W. Nicoll and S[tephen] Prudhoe. pp.58. [617.]

lxxxii. 1945. By S. Prudhoe. pp.163. [1261.]

lxxxiii. 1946. pp.93. [807.]

lxxxiv. 1947. pp.108. [859.]

lxxxv. 1948. pp.109. [847.]

lxxxvi. 1949. pp.108. [911.]

lxxxvii. 1950. pp.118. [1053.]

lxxxviii. 1951. pp.111. [1011.]

lxxxix. 1952. [*none.*]

xc. 1953. pp.115. [1160.]

xci. 1954. pp.126. [1100.]

xcii. 1955. pp.120. [1000.]

xciii. 1956. pp.118. [1100.]

xciv. 1957. pp.137. [1200.]

xcv. 1958. pp.131. [1300.]

xcvi. 1959. pp.84. [1300.]

xcvii. 1960. pp.87. [1500.]

xcviii. 1961. By S. Prudhoe, R. W. Sims, J. O. Malcolm. pp.114. [1400.]

in progress; the earlier issues were not published separately; the issues for 1906–1914 also form vols.vi–xiv of the International catalogue of scientific literature, N. Zoology.

ZOOLOGISCHER jahresbericht. . . . Vermes. Zoologische station zu Neapel: Berlin.

Animals

1885. Referenten: L. Örley, E[mil] v. Marenzeller, A. Lang. pp.93. [250.]

1886. Referenten: F. Zschokke, W. Kükenthal. pp.47. [175.]

1887. pp.75. [250.]

1888. Referenten: F. Zschokke, H. Eisig. pp.73. [300.]

1889. pp.65. [250.]

1890. Referenten: Th[eodor] Pintner, H. Eisig. pp.58. [300.]

1891. pp.75. [350.]

1892. pp.66. [400.]

1893. pp.64. [300.]

1894. pp.48. [300.]

1895. pp.61. [300.]

1896. pp.63. [300.]

1897. pp.64. [300.]

1898. pp.57. [300.]

1899. pp.64. [400.]

1900. pp.70. [300.]

1901. pp.70. [350.]

1902. pp.81. [400.]

1903. pp.81. [350.]

1904. pp.81. [400.]

1905. pp.70. [400.]

1906. pp.103. [500.]

1907. pp.87. [400.]

1908. pp.91. [400.]

1909. pp.81. [350.]

1910. Referenten: Max Rauther, H. Eisig. pp.80. [350.]

1911. pp.88. [400.]

1912. pp.90. [450.]

earlier issues and a later issue were not published separately.

F. ZSCHOKKE, Parasitische würmer. Central-kommission für schweizerische landeskunde: Bibliographie der schweizerischen landeskunde (section IV. 6. 8): Bern 1902. pp.xii.39. [600.]

Vertebrata.

ZOOLOGISCHER jahresbericht . . . [1880–1885: IV. Abtheilung: Vertebrata (1883–1885: Tunicata, vertebrata)]. Zoologische station zu Neapel: Leipzig.

1880. Redigirt von J. Vict. Carus. pp.iv.294. [1500.]

1881. pp.vi.314. [1500.]

1882. pp.iv.304. [1500.]

1883. Redigirt von Paul Mayer. pp.iv.334. [1500.]

1884. pp.iv.414. [1250.]

1885. pp.iv.336. [1500.]

1886. Referenten: M. v. Davidoff, C. Emery. pp.170. [700.]

Animals

1887. pp.180. [900.]

1888. Referenten: M. v. Davidoff, C. Emery, N. Löwenthal. pp.197. [900.]

1889. pp.182. [1000.]

1890. pp.205. [900.]

1891. Referenten: M. v. Davidoff [and others]. pp.221. [1000.]

1892. pp.248. [1100.]

1893. pp.221. [1000.]

1894. pp.250. [1000.]

1895. pp.250. [1000.]

1896. pp.230. [900.]

1897. pp.255. [900.]

1898. pp.231. [1000.]

1899. pp.214. [1000.]

1900. pp.210. [1100.]

1901. pp.219. [1000.]

1902. pp.236. [1100.]

1903. pp.254. [1100.]

1904. pp.271. [1300.]

1905. pp.254. [1100.]

1906. pp.251. [1200.]

1907. pp.266. [1200.]

1908. pp.235. [1200.]

1909. pp.274. [1100.]

1910. pp.266. [1000.]

1911. pp.262. [1100.]

1912. pp.251. [1000.]

earlier issues and a later issue were not published separately.

ARTHUR SMITH WOODWARD and CHARLES DAVIES SHERBORN, A catalogue of british fossil vertebrata. 1890. pp.xxxv.396. [9000.]

OLIVER PERRY HAY, Bibliography and catalogue of the fossil vertebrata of north America. Geological survey: Bulletin (no.179): Washington 1901. pp.868. [3000.]
—— Second bibliography [&c.]. Carnegie institution: Publication (no.390): Washington 1929–1930. pp.viii.916+xiv.1074. [10,000.]

THE ZOOLOGICAL RECORD. XIV. Vertebrata (general) i.e., complementary to mammalia, aves, reptilia, and pisces.
 xliii. 1906. By D. Sharp. pp.8. [75.]
 xliv. 1907. pp.12. [140.]
 xlv. 1908. pp.13. [145.]
 xlvi. 1909. pp.9. [89.]
 xlvii. 1910. pp.8. [85.]
 xlviii. 1911. pp.10. [117.]
 xlix. 1912. pp.10. [117.]
 l. 1913. pp.8. [88.]
 li. 1914. pp.4. [44.]
 lii. 1915. pp.4. [33.]

liii. 1916. pp.4. [26.]
liv. 1917. pp.3. [21.]
lv. 1918. pp.3. [26.]
lvi. 1919. pp.2. [7.]

the earlier and later issues were not published as separate sections; the issues for 1906–1914 also form part of vols.vi–xiv of the International catalogue of scientific literature, N. Zoology.

CASEY A[LBERT] WOOD, An introduction to the literature of vertebrate zoology, based chiefly on the titles in the Blacker library of zoology, the Emma Shearer Wood library of ornithology, the Bibliotheca osleriana and other libraries of the McGill university, Montreal. 1931. pp.xix.643. [15,000.]

BIBLIOGRAPHY of fossil vertebrates. Geological society of America: Special papers (no.27 &c.): [Washington].

    1928–1933. By C[harles] L[ouis] Camp and
      V[erstren] L[aurence] Vanderhoof. . . .
      (no.27): 1940. pp.[v].503. [7500.]
    1934–1938. By C. L. Camp, D[avid] [Natha-
      niel] Taylor and S[amuel] P[aul] Welles....
      (no.42): 1942. pp.v.663. [10,000.]
    1939–1943. By C. L. Camp, S. P. Welles

and Morton Green. Geologic society of America: Memoir (no.37): 1949. pp.v.371. [5000.]

1944–1948. . . . (no.57): 1953. pp.[v].465. [6000.]

1949–1953. By C. L. Camp and H. J. Allison. . . . (no.84): 1961. pp.xxxviii.532. [7000.]★

1954–1958. By C. L. Camp, H. J. Allison and R. H. Nichols. . . . (no.92): 1964. pp.xxvii. 647. [5000.]★

BIBLIOGRAPHY of vertebrate paleontology and related subjects. Society of vertebrate paleontology: [Chicago] 1946 &c.★

in progress; previously formed part of the society's News bulletin.

ALFRED SHERWOOD ROMER [*and others*], Bibliography of fossil vertebrates exclusive of north America, 1509–1927. Geological society of America: Memoir (no.87): New York 1962. pp.lxxxix. 772+[vi].773–1544. [35,000.]

Whales, whaling.

A COLLECTION of books, pamphlets, log books, pictures, etc. illustrating the whale fishery contained in the Free public library. New Bedford, Mass. 1907. pp.13. [600.]

Animals

— Second edition. A collection . . . illustrating whales and whale fishery. 1920. pp.24. [750.]

LIST of references on the whale industry. Library of Congress: Washington 1916. ff.4. [47.]*
— [supplement]. List of references on the whaling industry. 1935. ff.10. [123.]*

RECENT references to steam whalers. Science library: Bibliographical series (no.132): 1934. single sheet. [14.]*

TORBJØRN PEDERSEN and JOHAN T[IDEMAND] RUUD, A bibliography of whales and whaling. Select papers from the norwegian research work 1860–1945. Norske videnskaps-akademi: Hvalrådets skrifter (no.30): Oslo 1946. pp.32. [350.]

ARNE ODD JOHNSEN, Norwegian patents relating to whaling and the whaling industry. Kommandør Chr. Christensens hvalfangstmuseum: Publikasjon (no.16): Oslo 1947. pp.212. [533.]

ASBJØRN FJELD-ANDERSEN, Oversiktskatalog 1 over bøker, periodiske skrifter og særtrykk ved Hvalfangstmuseets bibliotek. Sandefjord 1961. pp.[ii].v.264. [4000.]

Xenopus laevis.

H[ARRY] ZWARENSTEIN, N[ORMAN] SAPEIKA and

H[ILLEL] A[BBE] SHAPIRO, Xenopus laevis. A bibliography. University of Capetown: Cape Town 1946. pp.[vii].51. [310.]

Zoocecideae.

FRIEDRICH AUGUST WILHELM THOMAS, Die Zoocecidien. . . . 1. band, 1ste lieferung. Verzeichnis der schriften über deutsche zoocecidien und cecidozoen bis einschliesslich 1906. Zoologica (vol.xxiv, no.61): Stuttgart 1911. pp.104. [1400.]

Plants

Algae.

THE CHEMICAL composition of algae. Science library: Bibliographical séries (no.21): [1931]. pp.[i].4. [85.]*

E[LMER] YALE DAWSON, A guide to the literature and distributions of the marine algae of the Pacific coast of north America. [Los Angeles 1946]. pp.134. [1250.]

SELECTED bibliography on algae. Nova Scotia research foundation: Halifax 1952. pp. 74. [1000.]
— Number [sic, edition] four. 1958. pp.109. [1500.]

SOME references to carbon dioxide/oxygen relations in algae (1941–1957). Science library: Bibliographical series (no.764): [1958]. f.1. [20.]*

Amla.

AMLA (emblica officinalis, Gaertn. syn.: phyllanthus emblica, L.)—chemical work. Pakistan national scientific and technical documentation centre: Pansdoc bibliography (no.265): Karachi 1960. pp.[ii].5. [50.]*

Plants

Anacardiaceae.

W. N. EDWARDS and F. M. WONNACOTT, Anacardiaceae. Fossilium catalogus (II.20): 's-Gravenhage 1935. pp.73. [1000.]

Apple.

W. H. RAGAN, Nomenclature of the apple. A catalogue of the known varieties referred to in american publications from 1804 to 1904. Department of agriculture: Bureau of plant industry: Bulletin (no.56): Washington 1905. pp.383. [17,500.]

Asclepiadaceae.

A[NNA] GERALDINE WHITING, A summary of the literature on milkweeds (asclepias spp.) and their utilization. Department of agriculture: Bibliographical bulletin (no.2): Washington 1943. pp.[ii].41. [95.]

Betulaceae.

K. NAGEL, Betulaceae. Fossilium catalogus (II. 8): Berlin 1916. pp.177. [2500.]

Bracken.

KENNETH WILLIAM BRAID, Bracken. A review of the literature. Commonwealth agricultural bureaux: Commonwealth publication (no.3): Farnham Royal 1959. pp.69.*

Plants

Bryology.

LEO F[RANCIS] KOCH, Bryology. SL bibliography series: New York [1961]. pp.[vi].20. [350.]*

Centrospermae.

JOHN W. SCHERMERHORN and MAYNARD W. QUIMBY, The [Eldin Verne] Lynn index. A bibliography of phytochemistry. Monograph 1. Order, *centrospermae*. Massachusetts college of pharmacy: Boston 1957. pp.46. [400.]

Charophyta.

J. GROVES, Charophyta. Fossilium catalogus (II. 9): Berlin 1933. pp.74. [600.]

Coconut.

LIST of references on cocoanuts and cocoanut oil. Library of Congress: Washington 1916. ff.3. [28.]*

S. R. SRINIVASA IYENGAR [ṢRĪNĪVASA AIYAṄGĀR], List of references on coconut & arecanut. Agricultural college and research institute: Library service: Bibliographical series (no.3): [Madras] 1941. pp.[ii].10. [200.]*

F[RANK] E. PETERS, Bibliographie sur les aspects nutritifs de la noix de coco. Commission du Pacifique sud: Documentation technique (no.58): Nouméa 1954. pp.v.43. [181.]*

—— Document revu et corrigé (no.95): 1956. pp.v.49. [205.]*

Cola.

COLA acuminata. Botany and culture. Science library: Bibliographical series (no.187): 1935. ff.[i].3. [46.]*

Cornaceae.

F. KIRCHHEIMER, Umbelliflorae: cornaceae. Fossilium catalogus (II. 23): 's-Gravenhage 1938. pp. xxii.168. [2500.]

Cryptogams.

[VICTOR GODEFRIN], Catalogue des ouvrages légués par m. J[ean]–B[aptiste]–H[enri]–J[oseph] Desmazières à la ville. Bibliothèque de la ville: Lille 1867. pp.xii.140. [651.]

Daffodils.

GENETICS and cytology of daffodils. Science library: Bibliographical series (no.179): 1935. single sheet. [37.]*

— Supplement. . . . (no.787): [1963]. pp.4. [80.]*

Plants

Dicotyledones.

W. N. EDWARDS, Dicotyledones (Ligna). Fossilium catalogus (II.17): Berlin 1931. pp.96. [1250.]

Equisetales.

W[ILLEM JOSEPH] JONGMANS, Equisetales. Fossilium catalogus (II. 2–5, 7, 9, 11): Berlin 1914–1924. pp.[ii].831. [15,000.]

Equisetum.

THE CHEMISTRY of equisetum spp. Science library: Bibliographical series (no.404): 1938. ff.2. [34.]★

Ferns. *see* Filicales.

Fig.

IRA J[UDSON] CONDIT and JULIUS ENDERUD, A bibliography of the fig. University of California: Hilgardia (vol.25): Berkeley 1956. pp.663. [7000.]★

Filicales.

THOMAS MOORE, Index filicum: a synopsis . . . with synonymes, references, &c. 1857–1863. pp. 4.ix–clxii.396. [10,000.]

parts 1–20 only; no more published.

Plants

CARL [FREDERIK ALBERT] CHRISTENSEN, Index filicum sive enumeratio omnium generum specierumque filicum et hydropteridum ab anno 1753 ad finem anni 1905 descriptorum. Hafniae [1905–] 1906. pp.lx.744. [25,000.]

— Supplementum, 1906–1912. 1913. pp.[iv]. 132. [4000.]

— Supplément préliminaire pour les années 1913, 1914, 1915, 1916. 1917. pp.[iv].60. [2000.]

— Supplementum tertium pro annis 1917–1933. 1934. pp.219. [7500.]

MAURICE BROWN, Index to north american ferns. Orleans, Mass. 1938. pp.217. [3000.]

Flora. *see* **Botany.**

Fruit.
    1. Periodicals, 418.
    2. General, 419.

1. *Periodicals*

PERIODICALS examined by the Imperial bureau of fruit production. East Malling 1931. ff.[36]. [149.]*

ESTHER M[ARIE] COLVIN, List of periodicals containing prices and other statistical and economic information on fruits, vegetables and nuts. Department of agriculture: Bureau of agricultural economics: Agricultural economics bibliography (no.55): Washington 1935. pp.[ii].238. [127.]★

2. *General*

[FRANÇOIS] THIÉRION, Revue bibliographique des principaux ouvrages français où il est traité de la taille des arbres fruitiers et particulièrement du pêcher. Troyes 1843. pp.108. [100.]

LIST of references on the California fruit growers' exchange. Library of Congress: Washington 1921. ff.2. [23.]★

KATHARINE G. RICE, Bibliography on the preservation of fruits and vegetables in transit and storage, with annotations. Department of agriculture: Library: Bibliographical contributions (no.4): Washington 1922. pp.[iii].76. [750.]★

FRUIT cultivation in deserts other than in the United States. Science library: Bibliographical series (no.56): [1932]. single sheet. [9.]★

MAMIE I[DELLA] HERB, Consumption of fruits and vegetables in the United States. An index to some sources of statistics. Department of agriculture: Bureau of agricultural economics: Agricultural economic bibliography (no.56): Washington 1935. pp.[ii].ii.125. [198.]★

WINTER washes for fruit trees with special reference to tar distillates. Science library: Bibliographical series (no.270): 1936. ff.12. [166.]★

G. K. ARGLES, A review of the literature on stock-scion incompatibility in fruit trees, with particular reference to pome and stone fruits. Imperial bureau of fruit production: Technical communication (no.9): East Malling 1937. pp.115. [194.]

THE DRYING of fruits and vegetables. References covering the period 1930–1937. Science library: Bibliographical series (no.380): 1938. ff.7. [115.]★
— 1930–1942 . . . (no.587): 1943. ff.10. [200.]★

IRRIGATION of small fruits. Science library: Bibliographical series (no.476): 1939. ff.3. [50.]★

VEGETABLE and fruit-growing. A small selection of the recently published books. Public libraries: Manchester 1944. pp.[6] (folder). [150.]

M[ARY] G. SMITH, Deciduous fruit. A bibliography. University of Cape Town: School of librarianship: Bibliographical series: [Capetown] 1947. ff.25. [168.]★
limited to south african fruit.

A[LEKSANDR] E[VGENEVICH] CHZHAO, Украсим родину садами. Рекомендательный указатель литературы. Государственная...библиотека СССР им. В. И. Ленина: Москва 1956. pp.31. [40.]

VIKTORIYA ANDREEVNA GALUNSKAYA, За расширение садоводства в северо-западной зоне. Рекомендательный список литературы. Государственная... публичная библиотека имени М. Е. Салтыкова-Щедрина: Ленинград 1957. ff.20. [100.]*

ROBERT LANIER KNIGHT and ELIZABETH KEEP, Abstract bibliography of fruit breeding and genetics. Rubus and ribes a survey. Commonwealth bureau of horticulture and plantation crops: Technical communication (no.25): East Malling 1958. pp.254.

DAGMAR FRANCISCI, O pěstování zeleniny v zahrádkách. Universita: Knihovna: Výběrový seznam (no.32): Brně 1959. pp.[iv].26. [60.]*

Hydnocarpus.

CULTIVATION of hydnocarpus spp. for the production of chaulmoogra oil. Science library: Bibliographical series (no.449): 1939. ff.2. [24.]*

Iris.

TRESSIE COOK, Louisiana irises. A bibliography.

Southwestern Louisiana institute: Lafayette 1953. pp.16. [200.]

Juglandaceae.

K. NAGEL, Juglandaceae. Fossilium catalogus (II. 6): Berlin 1915. pp.87. [1500.]

Lichenaceae.

AUGUST VON KREMPELHUBER, Geschichte und litteratur der lichenologie von den ältesten zeiten bis zum schlusse des jahres 1865. München 1867–1869. pp.xiv.616+viii.776. [1412.]

—— III. band. Die fortschritte und die litteratur der lichenologie in dem zeitraume von 1866–1870 incl. nebst nachträgen zu den früheren perioden. 1872. pp.xvi.261. [450.]

G[USTAV] LINDAU and P[AUL] SYDOW, Thesaurus litteraturae mycologicae et lichenologicae. Lipsiis [1907] 1908–1917. pp.vii.903+[ii].808+iv.766+xiii.609+viii.527. [42,000.]

ALEXANDER WILLIAM EVANS and ROSE MEYROWITC, Catalogue of the lichens of Connecticut. State geological and natural history survey: Bulletin (no.37): Hartford 1926. pp.49.[vii]. [500.]

COMPOSITION of lichens. Science library: Bibliographical series (no.164): 1934. ff.[i].6. [111.]*

— Supplement. . . . (no.758): [1958]. pp.11. [150.]*

Plants

Lycopodiales.

W[ILLEM JOSEPH] JONGMANS, Lycopodiales. Fossilium catalogus (II. 1, 15, 16, 18, 21, 22): Berlin ['s-Gravenhage] 1913–1937. pp.ciii.1331. [25,000.]

Malvales.

JOHN W. SCHERMERHORN and MAYNARD W. QUIMBY, The [Eldin Verne] Lynn index. A bibliography of phytochemistry. Monograph II. Order, *malvales*. Massachusetts college of pharmacy: Boston 1958. pp.39. [450.]

Marasmius.

OCCURRENCE and habitat of marasmius oreades, one of the fungi causing fairy rings. Science library: Bibliographical series (no.257): 1935 [1936]. single sheet. [8.]

Mistletoes.

R. G. SANZEN-BAKER, Literature on the mistletoes. Imperial forestry institute: Institute paper (no.12): Oxford 1938. pp.iii.14. [93.]*

Morus.

[P. A. BOISSIER SAUVAGES DE LA CROIX], Catalogue des auteurs qui ont écrit sur les vers à soie & sur les mûriers. [Nîmes 1763]. pp.8. [100.]

ROBERTO DI TOCCO, Bibliografia del filugello (bombyx mori L.) e del gelso (morus alba L.). Ente nazionale serico: Milano 1927. pp.xliii.262. [3850.]

Moss. *see* **Musci** *and* **Sphagna.**

Mulberry. *see* **Morus.**

Musci.

E. G. PARIS, Index bryologicus sive enumeratio muscorum hucusque cognitorum adjunctis synonyma distributioneque geographica locupletissimis. Parisiis 1894–1898. pp.[ii].vi.1380. [40,000.]
— — Editio secunda. 1904 &c.

AUGUST JAEGER and FR. SAUERBECK, Genera et species muscorum systematica disposita, seu adumbratio floræ muscorum totius orbis terrarum. Sancti Galli [1897]. pp.[ii].245–299+[ii]. 357–451 + [ii].309–490 + [ii].61–236 + [ii].53– 278 + [ii].85–188 + [ii].201–371 + [ii].211–454 +[ii].257–514.213–252. [25,000.]

CARRAGEEN, Irish moss. Science library: Bibliographical series (no.98): [1933]. ff.[i].2. [24.]★
— 2nd edition. . . . (no.416): 1938. ff.3. [42.]★
— — Supplement. . . . (no.752): [1957]. ff.5. [100.]★

Plants

GENEVA SAYRE, Dates of publications describing musci, 1801–1827. Troy, N.Y. 1959. pp.[v].102. [350.]*

Muscineae.

H. N. DIXON, Muscineae. Fossilium catalogus (I. 13): Berlin 1927. pp.116. [1250.]

Mushrooms.

JOSEPHINE A[DELAIDE] CLARK, Reference list of publications relating to edible and poisonous mushrooms. Department of agriculture: Library bulletin (no.20): Washington 1898. pp.16. [300.]

SELECT list of works relating to edible and poisonous mushrooms. Library of Congress: Washington 1913. ff.3. [20.]*

— Additional references. 1928. ff.2. [27.]*

Mycology. [*see also* Botany: Pathology.]

W[ILLIAM] G[ILSON] FARLOW and WILLIAM TRELEASE, A list of works on north american fungi. Library of Harvard university: Bibliographical contributions (no.25): Cambridge, Mss. 1887. pp.36. [654.]

— — A supplementary list. . . . By W. G. Farlow . . . (no.31): 1888. pp.9. [100.]

Plants

WILLIAM G. FARLOW, Bibliographical index of north american fungi. . . . Vol.I. — Part I. Abrothallus to badhamia. Carnegie institution: Publication (no.8): Washington 1905. pp.xxxv.312. [12,500.]

no more published.

G[USTAV] LINDAU and P[AUL] SYDOW, Thesaurus litteraturae mycologicae et lichenologicae. Lipsiis [1907–]1908–1917. pp.vii.903+[ii].808+iv.766+xiii.609+viii.527. [42,000.]

ÉMILE JAHANDIEZ, Essai de bibliographie mycologique varoise. Draguignan [printed] 1912. pp.3. [15.]

THE REVIEW of applied mycology. Imperial [*afterwards:* Commonwealth] bureau of mycology: Kew.

    i. [Edited by E. J. Butler]. 1922. pp.iii.502. [500.]

    ii. 1923. pp.iii.654. [600.]

    iii. 1924. pp.iii.848. [800.]

    iv. 1925. pp.[iii].868. [1250.]

    v. 1926. pp.[iii].874. [1500.]

    vi. 1927. pp.[iii].880. [1500.]

    vii. 1928. pp.[iii].928. [1750.]

    viii. 1929. pp.[iii].942. [1750.]

    ix. 1930. pp.[iii].939. [1750.]

x. 1931. pp.[iii].939. [1750.]

xi. 1932. pp.[iii].938. [1750.]

xii. 1933. pp.[iii].923. [1750.]

xiii. 1934. pp.[iii].938. [1750.]

xiv. [Edited by Samuel Paul Wiltshire]. 1935. pp.[iii].950. [1750.]

xv. 1936. pp.[iii].966. [1750.]

xvi. 1937. pp.[iii].979. [1750.]

xvii. 1938. pp.[iii].979. [1750.]

xviii. 1939. pp.[iii].943. [1750.]

xix. 1940. pp.[iii].834. [1500.]

xx. 1941. pp.[iii].703. [1250.]

xxi. 1942. pp.[iii].605. [1000.]

xxii. 1943. pp.[iii].571. [1000.]

xxiii. 1944. pp.[iii].572. [1000.]

xxiv. 1945. pp.[iii].574. [1000.]

xxv. 1946. pp.vii.671. [1250.]

xxvi. 1947. pp.[iii].652. [1250.]

xxvii. 1948. pp.[iii].674. [1250.]

xxviii. 1949. pp.[iii].742. [1500.]

xxix. 1950. pp.[iii].734. [1500.]

xxx. 1951. pp.[ii].738. [1500.]

xxxi. 1952. pp.[iii].743. [1500.]

xxxii. 1953. pp.[iv].816. [1500.]

xxxiii. 1954. pp.[iv].896. [2000.]

LOUIS C[HARLES] C[HRISTOPHER] KRIEGER, Catalogue of the mycological library of Howard

Plants

A[twood] Kelly. Baltimore 1924. pp.[v].ix.260. [6500.]

privately printed.

OCCURRENCE and habitat of marasmius oreades' one of the fungi causing fairy rings. Science library: Bibliographical series (no.257): 1935 [1936]. single sheet. [8.]★

CATALOGUS van de bibliotheek der Landbouw-hoogeschool. Bruikleen van de Nederlandsche mycologische vereeniging. Wageningen 1936. ff.[iii].57.[vi]. [450.]★

— 1ste supplement. [1937]. pp.6. [50.]★

AN ANNOTATED bibliography of medical mycology. Imperial [Commonwealth] mycological institute: Kew.

    1943. Edited by S[amuel] P[aul] Wiltshire. pp.32. [218.]

    1944. pp.33–74. [202.]

    1945. pp.75–122. [225.]

    1946. pp.123–174. [200.]

    1947. pp.175–227. [206.]

    1948. pp.229–292. [231.]

    1949. pp.293–362. [272.]

    1950. pp.363–429. [256.]

    1951. pp.431–478. [222.]

        [continued as:]

Plants

Review of medical and veterinary mycology.
  i. 1952. pp.479–590. [517.]
  ii. 1953–1957. Edited by S. P. Wiltshire,
    J. C. F. Hopkins. pp.[iv].631. [2464.]
  iii. 1958–1960. pp.[iv].400. [1918.]
  iv. 1961–1963. pp.[iv].480. [2224.]
in progress.

FRANCIS D. HORIGAN and CARY R. SAGE, Fungal
growth on painted surfaces: a bibliography.
Quartermaster research and development laboratories: Technical library: Bibliographic series
(no.1): [Washington] 1947. pp.iii.33. [130.]*

FRANCIS D. HORIGAN and CARY R. SAGE, The
mycology of leather. A bibliography. QM
research & development laboratories: Technical
library: Bibliographic series (no.4): Philadelphia
[1948]. ff.[iii].37.viii. [115.]*

FRANCIS D. HORIGAN and CARY R. SAGE, The
technology of fungi, mold, and mildew. A
literature and patent survey. QM research &
development laboratories: Technical library: Philadelphia 1950. pp.159. [1585.]*

GEORGE WILLIAM FISCHER, The smut fungi. A
guide to the literature, with bibliography. New
York [1951]. pp.x.387. [3353.]

BIBLIOGRAPHY of systematic mycology. Commonwealth mycological institute: Kew.*

1957. [Edited by G. M. Waterhouse]. pp.[ii]. 41. [832.]

1958. pp.[ii].50. [903.]

RAFAELE CIFERRI and PIERO REDAELLI, Bibliographia mycopathologica (1800–1940). Biblioteca bibliografica italica (vols.18, 20): Firenze 1958. pp.3–410+3–402. [14,506.]

DOROTHY BOCKER, Fungus infections. A bibliography . . . 1952 through September 1958. National library of medicine: Reference division: Washington 1959. pp.[v].90. [942.]*

APPLIED mycology. A classified bibliography of philippine publications on fungi and their role. National institute of science and technology: Division of documentation: Series of philippine scientific bibliographies (no.1): Manila 1960. pp.14. [288.]*

Myxomycetes.

WILLIAM C[ODMAN] STURGIS, A guide to the botanical literature of the myxomycetes from 1875 to 1912. Colorado college publications: Science series (vol.xii, no.1): Colorado Springs 1912. pp.[ii].385–433. [750.]

Plants

Nectaries.

HARRY DARROW BROWN, Nectar and nectar glands. SL bibliography series: New York [1961] pp.[vi].20. [450.]*

Orchids.

MARGARET WILSON, A bibliography of south african orchids. University of Cape Town: School of librarianship: Bibliographical series: [Capetown] 1957. pp.[iii].v.21. [182.]*

Paleobotany. *see* Botany: Paleontology.

Phanerogamae.

K[ARL] RICHTER, Plantæ Europeæ. Enumeratio systematica et synonymica plantarum phanerogamicarum in Europa sponte crescentium vel mere inquilinarum. [vol.ii: Emendavit ediditque M. Gürke]. 1890–1897[–1903]. pp.vii.378+vi.480. [20,000.]
incomplete; no more published.

B[ENJAMIN] DAYDON JACKSON, Index kewensis plantarum phanerogamarum, nomina et synonyma omnium generum et specierum a Linnaeo usque ad annum MDCCCLXXXV completens. . . . Ductu et consilio [sir] Josephi D. Hooker confecit B. Daydon Jackson. Oxonii [1893–]1895. pp.

431

xiv.1268+vii.1299. [425,000.]

— — Supplementum.

    i. 1886–1895. Confecerunt Theophilus Durand & B. Daydon Jackson. Bruxellis 1901–1906. pp.[v].519. [75,000.]

    ii. 1896–1900. Ductu et consilio W. T. Thiselton-Dyer confecerunt Herbarii Horti Regii Botanici Kewensis curatores. Oxonii 1904. pp.[iii].204. [30,000.]

    iii. 1901–1905. Ductu et consilio D[avid] Prain confecerunt . . . curatores. 1908. pp.[iii].193. [27,500.]

    iv. 1906–1910. 1913. pp.[iii].252. [37,500.]

    v. 1911–1915. 1921. pp.[iii].277. [40,000.]

    vi. 1916–1920. Ductu et consilio A. W. Hill confecerunt . . . curatores. 1926. pp.[iii].222. [32,500.]

    vii. 1921–1925. 1929. pp.[iii].260. [37,500.]

    viii. 1926–1930. 1933. pp.[iii].256. [37,500.]

    ix. 1931–1935. 1938. pp.[iii].305. [45,000.]

    x. 1936–1940. 1947. pp.[iii].251. [35,000.]

    xi. 1941–1950. 1953. pp.[iii].273. [40,000.]

Pine tree.

A[LLAN] P[RIESTLEY] THOMSON, Bibliography of Pinus radiata. New Zealand forest service: Forest research institute: Forest research notes (vol.i, no.1): Wellington 1950. ff.[i].14. [125.]★

DAVID TACKLE and D. I. CROSSLEY, Lodgepole pine bibliography. Forest service: Intermountain forest and range experiment station: Research paper (no.30): Ogden, Utah 1953. ff.57. [442.]

Plant. *see* **Botany.**

Poplar.

JOACHIM H. VOLKMANN, Bibliographie des internationalen pappelschrifttums. Nationale pappelkommission: Bonn 1954–1955. ff.[iv].xxvi.202+ [iii].203–345. [2120.]

Ramie.

LIST of references on ramie. Library of Congress: [Washington] 1921. ff.6. [67.]★

Redwood, tree.

BIBLIOGRAPHY of the redwoods. Reading-list of articles, pamphlets and books on the Sequoia sempervirens. Save-the-redwoods league: [Berkeley, Cal.] 1935. pp.16. [200.]

FRANCIS P[ELOUBET] FARQUHAR, Yosemite, the big trees and the high Sierra. A selective bibliography. Berkeley 1948. pp.xii.104. [150.]

EMANUEL FRITZ, California coast redwood (Sequoia sempervirens (D. Don) Endl.). An annotated bibliography. Foundation for american resource management: San Francisco 1957. pp. xv.267. [2003.]

Rice.

LIST of references on the rice industry. Library of Congress: Washington 1916. ff.2. [12.]*
— [another edition]. List of references on rice and the rice industry. 1920. ff.14. [146.]*

RICE breeding bibliography. Imperial bureau of plant genetics: Cambridge [1931]. ff.[ii].24. [125.]*

G[RAHAM] V[ERNON] JACKS and M[ARIAN] KATHLEEN MILNE, An annotated bibliography of rice soils and fertilizers. Commonwealth bureau of soil science [Harpenden] [and] Food and Agriculture organization of the United Nations: Rome 1954. pp.v.180. [952.]

LORENZO VANOSSI, Bibliografia del riso fina al 1899. Ente nazionale risi: Ufficio studi: [Milan 1959]. pp.47. [800.]

INTERNATIONAL bibliography of rice research. International rice research institute: New York 1963. pp.lvi.881. [7274.]*

Rose.

CL[AUDE] ANT[OINE] THORY, Bibliotheca botanica rosarum. Paris 1818. pp.18. [400.]

MARIANO VERGARA, Bibliografía de la rosa. Madrid 1892. pp.319. [600.]
printed on one side of the leaf.

LIST of references on legends and symbolism of the rose. Library of Congress: Washington 1924. ff.5. [49.]*

Salix.

THE HISTOLOGY, microchemistry and biochemistry of salix species. Science library: Bibliographical series (no.120): 1934. ff.[i].2. [25.]*

Sapindaceae.

W. N. EDWARDS and F. M. WONNACOTT, Sapindaceae. Fossilium catalogus (II. 4): Berlin 1928. pp.84. [1250.]

Sphagna.

JULES CARDOT, Répertoire sphagnologique. Catalogue alphabétique . . . du genre Sphagnum, avec la synonymie, la bibliographie. Autun [printed] 1897. pp.200. [3000.]

Plants

Sunflower.

MARJORIE F. WARNER, Sunflower (helianthus annuus). Department of agriculture: Library: Agricultural library notes (vol.v, no.1–3, supplement): Washington 1930. pp.[iii].18. [150.]*

Ulmaceae.

K. NAGALHARD, Ulmaceae. Fossilium catalogus (1. 10): Berlin 1922. pp.[iii].84. [1500.]

Special Subjects

Acids.

[MARGARET L. BUCH], A bibliography of organic acids in higher plants. Department of agriculture: Agricultural research service (no.73-18): Washington 1957. pp.136. [750.]

— — [another edition]. Department of agriculture: Agricultural research service: Agricultural handbook (no.164): [1960]. pp.100. [823.]

Animal mythology.

MARGARET W. ROBINSON, Fictitious beasts. A bibliography. Library association: 1961. pp.76. [349.]

Animals, protection of.

H. FISCHER-SIGWART, Tierschutz, Centralkommission für schweizerische landeskunde: Bibliographie der schweizerischen landeskunde (section V.9k): Bern 1906. pp.x.101. [2000.]

HUMANITARIANISM. A selected list of books. Third edition. [National book council:] Bibliography (no.57): 1929. pp.[2]. [60.]

A. M. SPEIGHT, Game reserves and game protection in Africa, with special reference to south Africa. A bibliography. University of Cape Town: School of librarianship: Bibliographical series: [Capetown] 1950. pp.iv.32. [231.]★

Bactericides.

BACTERICIDAL substances derived from micro-organisms. Annotated bibliography. Merck & co.: Rahway, N. J. [1941]. ff.[ii].25. [46.]★

— [another edition]. [1942]. ff.[ii].47 [*sic*,50]. [93.]★

— — Supplement. [1942]. ff.[ii].28. [56.]★

Bacteriology.

JAHRESBERICHT über die fortschritte in der lehre von den pathogenen mikroorganismen, umfassend bacterien, pilze und protozoën. Braunschweig [vols.xv–xxvii: Leipzig].

    i. 1885. Von P[aul] Baumgarten. 1886. pp.[viii].192. [200.]

    ii. 1886. 1887. pp.viii.458. [535.]

    iii. 1887. 1888. pp.viii.517. [818.]

    iv. 1888. 1889. pp.xi.387 [*sic*, 587]. [959.]

    v. 1889. 1890. pp.xi.632. [1017.]

    — Namen- und sach-register . . . 1885–1889. 1891. pp.[iii].98.

    vi. 1890. 1891. pp.xi.651. [961.]

    vii. 1891. 1893. pp.xi.919. [1269.]

    viii. 1892. 1894. pp.xi.807. [1337.]

    ix. 1893. Herausgegeben von P. Baumgarten und F. Roloff. 1895. pp.xi.856. [1491.]

x. 1894. Herausgegeben von P. Baumgarten und F. Tangl. 1896. pp.x.846. [1593.]

— Namen- und sach-register . . . 1885 bis 1894. Bearbeitet von B. Honsell und E. Ziemke. 1896. pp.[ii].280.

xi. 1895. 1897. pp.xi.794. [1685.]

xii. 1896. 1898. pp.xi.896. [1933.]

xiii. 1897. 1899. pp.xii.1063. [2223.]

xiv. 1898. 1900. pp.xii.1055. [2222.]

xv. 1899. 1901. pp.xii.1040. [2508.]

xvi. 1900. 1902. pp.xii.812. [1853.]

xvii. 1901. 1903. pp.xii.1114. [3151.]

xviii. 1902. 1904. pp.xii.1364. [3314.]

xix. 1903. 1905. pp.xii.1220. [1337.]

xx. 1904. 1906. pp.xii.1106. [3469.]

xxi. 1905. 1907. pp. xii. 941. [2861.]

xxii. 1906. 1908. pp.xii.964. [2623.]

xxiii. 1907. 1907. pp.xii.940. [3027.]

xxiv. 1908. Herausgegeben von P. von Baumgarten und Walter Dibbelt. 1911. pp.xii.1136. [3312.]

xxv. 1909. 1912. pp.xii.1159. [3484.]

xxvi. 1910. 1913. pp.xii.1482. [2729 (*sic*, 3729).]

xxvii. 1911. 1917. pp.xii.1156. [3566.]

no more published.

ALEXANDER RAMSEY, The scientific roll. . . .

Bacteria. [1900–]1905[–1913]. pp.viii.528+viii. 514+515–586. [2000.]

published as a periodical; incomplete; no more published.

BACTERIOLOGY. International catalogue of scientific literature (section R): Royal society.

    i. 1902. pp.xv.314. [2206.]
    ii. 1903. pp.viii.435. [3132.]
    iii. 1905. pp.viii.520. [3925.]
    iv. 1906. pp.viii.500. [3615.]
    v. 1907. pp.viii.837. [6208.]
    vi. 1909. pp.viii.1040. [9179.]
    vii. 1909. pp.viii.771. [6426.]
    viii. 1911. pp.viii. 636. [6545.]
    ix. 1912. pp.viii.530. [5679.]
    x. 1914. pp.viii.602.182.23. [10,750.]
    xi. 1915. pp.viii.583.148.27. [10,000.]
    xii. 1917. pp.viii.300.111.24. [4654.]
    xiii. 1918. pp.viii.413.138.24. [6362.]
    xiv. 1920. pp.viii.354.132.22. [4816.]

no more published; the 10th–14th issues contain a separate section QR: Serum physiology.

THE REVIEW of bacteriology [*afterwards:* protozoology] and general parasitology: an epitome of general parasitology, bacteriology, and allied subjects in their relationship to pathology and hygiene.

i. 1911. Edited by Alexander G. R. Foulerton. pp.108. [341.]

ii. 1912. Edited by A. G. R. Foulerton and Charles Slater. pp.xviii.127. [481.]

iii. 1913. pp.xix.120. [529.]

iv. 1914. pp.xi.78. [315.]

v. 1915. pp.xvi.80. [333.]

vi. 1916. pp.xii.87. [320.]

vii. 1917. pp.xv.157. [350.]

viii. 1918. pp.xvi.148. [365.]

ix. 1919. pp.xvi.132. [353.]

no more published.

ABSTRACTS of bacteriology. Baltimore.

i. 1917. pp.598. [1622.]

ii. 1918. Editor: A. Parker Hitchens. pp.[ii]. 426. [2150.]

iii. 1919. pp.[ii].445. [2146.]

iv. 1920. pp.[ii].477. [2328.]

v. 1921. pp.[ii].602. [2606.]

vi. 1922. pp.[ii].544. [2120.]

vii. 1923. pp.[ii].508. [2233.]

viii. 1924. pp.[ii].462. [2331.]

ix. 1925. pp. [ii].498. [2296.]

continued as part of Biological abstracts.

F[ELIX] LÖHNIS, Studies upon the life cyclus of the bacteria. . . . 1. Review of the literature, 1838–1918. National academy of sciences:

Memoirs (vol.xvi, no.2): Washington [1922]. pp.335. [1350.]

EFFECT of ultra short radio waves on bacteria. Science library: Bibliographical series (no.177): 1935. single sheet. [13.]★

— [second edition]. Action of ultra-short radio waves on bacteria and insects . . . (no.257 [*sic*, 259]): 1936. ff.3. [36.]★

— — Supplement . . . (no.666): 1949. ff.2. [27.]★

SELECT list of books on bacteriology and its industrial applications exclusive of medicine, 1930 to date. Science library: Bibliographical series (no.275): 1936. ff.2. [35.]★

ELIZABETH [FLORENCE] MCCOY and L[ELAND] S[WINT] MCCLURG, The anaerobic bacteria and their activities in nature and disease. A subject bibliography. Berkeley, Cal.1939. pp.xxiii.295 +xi.602. [10,500.]★

— — Supplement one, literature for 1938 and 1939. 1941. pp.xxiii.244. [6000.]★
no more published.

ACTION of high frequency fields on insects and bacteria. Science library: Bibliographical series (no.676): 1949. ff.2. [35.]★

Special Subjects

HANS SCHMIDT, Bacteriology and immunology. Office of military government for Germany: Field information agencies technical: Fiat review of german science, 1939–1946: Wiesbaden 1947. pp.[xi].155. [1250.]
the text is in german.

BACTERIOLOGICAL warfare: a list of references in Soviet publications, 1929 to February 1952. Library of Congress: Washington 1952. ff.11. [78.]*

CATALOGUE of publications on anaerobic bacteria in the library of the Veterinary institute, Bogor (Indonesia) collected and donated . . . by F. C. Kraneveld. Bogor 1954. pp.194. [4000.]*

FREDA and PETER GRAY, Annotated bibliography of works in latin alphabet languages on biological microtechnique. Dubuque [1956]. pp.viii.116. [750.]

[R. TRAUTMANN, V. BEDNÁŘOVA and O. LÍBAL], Hygiena, epidemologie a bakteriologie. Universita: Knihovna: Výběrový seznam (no.15): Brně 1957. ff.[vi].98. [900.]*

HANSJÜRGEN RAETIG, Bakteriophagie 1917 bis 1956, zugleich ein vorschlag zur dokumentation wissenschaftlicher literatur. Stuttgart 1958. pp. xix.215+[iv].344. [5655.]

THOMAS H. GRAINGER, A guide to the history of bacteriology. Chronica botanica (no.18): New York [1958]. pp.xi.210. [2802.]

HILMER A. FRANK and L. LEON CAMPBELL, A bibliography of bacterial spores. State college of Washington: Washington agricultural experiment stations: Stations circular (no.323): [Pullman] 1958. pp.ii.117. [1250.]*

ESTHER M[ARTIA] SCHLUNDT and ANN E. KERKER, The reference library in microbiology. A selected list of titles. Purdue university: Libraries: Lafayette 1959. ff.[v].64. [900.]*

Biochemistry.

ANNUAL review of biochemistry. Stanford university.

    i. 1932. pp.vii.724. [8000.]
    ii. 1933. pp.vii.564. [5500.]
    iii. 1934. pp.viii.558. [5500.]
    iv. 1935. pp.vii.639. [7000.]
    v. 1936. pp.ix.640. [7000.]
    vi. 1937. pp.ix.708. [7500.]
    vii. James Murray Luck, editor. 1938. pp.ix.
       571. [5500.]
    viii. 1939. pp.ix.676. [7000.]
    ix. 1940. pp.ix.744. [8000.]

x. 1941. pp.ix.692. [7000.]

xi. 1942. pp.ix.736. [8000.]

xii. 1943. pp.ix.704. [7500.]

xiii. 1944. pp.ix.795. [8500.]

xiv. 1945. pp.x.856. [9000.]

xv. 1946. pp.viii.687. [7500.]

xvi. 1947. pp.xi.740. [8000.]

xvii. 1948. pp.ix.801. [8500.]

xviii. 1949. pp.ix.739. [8000.]

xix. 1950. pp.xi.596. [6500.]

xx. 1951. pp.ix.648. [7000.]

— Cumulated index to vol. 11 to 20. Edited by A. C. Griffin. 1952. pp.iv.377.

xxi. 1952. pp.ix.781. [8500.]

xxii. 1953. pp.x.729. [.]

xxiii. 1954. pp.x.636. [.]

xxiv. 1955. pp.xvi.805. [.]

no more published.

FELIX HAUROWITZ, Biochemie des menschen und der tiere seit 1914. Wissenschaftliche forschungsberichte: Naturwissenschaftliche reihe (vol.xii): Dresden &c. 1925. pp.xii.148. [1100.]

— — II. Teil (1924–1931). . . . (vol.xxvi): 1932. pp.x.152. [1250.]

ABSTRACTS of chemical papers. . . . A 3. Biochemistry. Bureau of chemical abstracts: 1937 &c.

details of this work are entered under Chemistry: General, below.

RICHARD KUHN [*and others*], Biochemistry. Office of military government for Germany: Field information agencies technical: Fiat review of german science 1939–1946: Wiesbaden 1947–1948. pp.[xv].218+[iv].241+[iv].201+[iv].318. [3000.] *the text is in german.*

РЕФЕРАТИВНЫЙ журнал. Химия-биологическая химия. Академия наук СССР: Институт научной информации: Москва.

1955. pp.80.vii + 84.vii + 77.vii + 71.viii + 90.viii + 82.viii + 96.viii + 95.ix + 105.ix + 109.ix + 109.x + 106.x + 89.ix + 89.ix + 97.x + 103.x + 103.x + 92.x. + 98.x + 104.xii + 110.xi + 111.xii + 114.xiv + 640. [18,004.]

1956. pp.222.vii + 109.vi + 130.vii + 118.vii + 116.viii + 129 + 133 + 117 + 104 + 109 + 125 + 117 + 136 + 105 + 112 + 112 + 100 + 116 + 120 + 120 + 128 + 128 + 141. [23,973.]

1957. pp.136 + 136 + 132 + 133 + 133 + 144 + 133 + 145 + 140 + 161 + 141 + 128 + 120 + 137 + 137 + 137 + 136 + 132 + 125 + 125 + 100 + 105 + 101 +

104. [27,023.]

1958. pp.160 + 152 + 148 + 152 + 152 +
156 + 152 + 148 + 120 + 160 + 160 +
168 + 160 + 168 + 160 + 148 + 141 +
164 + 167 + 160 + 157 + 183 + 201 +
197. [33,207.]

1959. pp.137 + 125 + 149 + 183 + 181 +
196 + 181 + 189 + 196 + 188 + 187 +
193 + 188 + 201 + 204 + 197 + 193 +
205 + 197 + 169 + 168 + 168. [32,818.]

1960. pp.169 + 200 + 201 + 200 + 196 +
209 + 209 + 224 + 201 + 201 + 232 +
208 + 228 + 208 + 200 + 208 + 197 +
180 + 188 + 185 + 188 + 181 + 188 +
208 + 180. [35,182.]

1961. pp.189 + 208 + 196 + 192 + 196 +
184 + 188 + 156 + 194 + 192 + 176 +
168 + 172 + 170 + 168 + 144 + 160 +
153 + 164 + 148 + 144 + 200 + 148 +
200. [37,500.]

1962. pp.168 + 168 + 136 + 176 + 208 +
196 + 192 + 188 + 200 + 212 + 200 +
200 + 200 + 188 + 200 + 200 + 196 +
188 + 168 + 168 + 188 + 212 + 216 +
280. [40,000.]

1963. pp.196 + 156 + 183 + 179 + 184 +
188 + 232 + 216 + 212 + 236 + 214 +

264 + 236 + 196 + 120 + 136 + 164 +
204 + 211 + 220 + 272 + 288 + 156 +
156. [40,000.]

*in progress; previously formed part of the sections
devoted to chemistry and biology.*

E. E. GONCHAROVA, N. M. POLYAKOVA and TS. M.
SHTUTMAN, Биохимия нервной системы. Библиографический указатель отечественной
литературы 1868–1954. Академия наук Ук-
раинской ССР: Институт биохимии: Киев
1957. pp.88. [1000.]

J. A. MCCORMICK, Isotopes in biochemistry and
biosynthesis of labeled compounds. A selected list
of references. United States atomic energy commission: Technical information service extension:
Oak Ridge, Tenn. 1958. pp.89. [2430.]

BIBLIOGRAPHY of papers published by László
Zechmeister and co-authors in the fields of chemistry and biochemistry, 1913–1958. Wien 1958.
pp.[vii].22. [257.]

JEAN ÉMILE COURTOIS, Biochemistry. French
bibliographical digest (no.27): New York 1959.
pp.171. [1200.]

JEAN ÉMILE COURTOIS, Biochimie. Association
pour la diffusion de la pensée française: Bibliographie française établie à l'intention des lecteurs

étrangers: 1960. pp.152. [1000.]
limited to works published in 1951–1956.

W[ILLIAM] H. FITZPATRICK and W[ILLIAM] HENRY
SEBRELL, Biochemistry of the USSR. A 25-year
literature search of the journal Biokhimiya. Israel
program for scientific translations: [Jerusalem
1961]. pp.[iii].199. [2000.]*

Biophysics.

BORIS RAJEWSKY, MICHAEL SCHÖN [*and others*],
Biophysics. Office of military government for
Germany: Field information agencies technical:
Fiat review of german science, 1939–1946: Wies-
baden 1948. pp.[viii].257+[iv].411. [2500.]
   the text is in german.

БИОФИЗИКА, биохимия, физиология, микробиология [генетика]. Систематический
указатель статей в иностранных журналах.
Всесоюзная государственная библиотека
иностранной литературы: Москва.

   1952. pp.52.52.52.56.57.76. [7500.]
   1953.   pp.136.72.76.72.81.68.60.69.65.68.
      [12,500.]
   1954. pp.68.76.104.92.97.89.89.92. [12,700.]
   1955.   pp.69.84.88.80.85.89.75.76.73.72.
      [15,000.]

1956.   pp.84.88.73.80.60.64.60.68.84.80.
[15,000.]

1957.

1958.   pp.112.82.100.84.81.116.108.112.129.
64. [20,000.]

1959. pp.109.104.108.84.81.

in progress?

Biotin.

BIOTIN. Annotated bibliography. Merck & co.:
Rahway, N.J. [1942]. ff.[ii].49. [100.]*

— Revised. 1944. ff.[ii].104. [300.]*

Bompland, Aimé.

HENRI CORDIER, Papiers inédits du naturaliste
Aimé Bompland conservés à Buenos Aires. 1910.
pp.25. [100.]

*the papers are at the Instituto de botánica y farma-
cología of the university.*

British museum.

[BERNARD BARHAM WOODWARD], List of works
relating to the Natural history departments of
the British museum. British museum (Natural
history): 1911. pp.71. [1000.]

Canary islands.

AGUSTÍN MILLARES CARLO, Ensayo de una bio-
bibliografía de escritores naturales de las islas

Canarias (siglos XVI, XVII y XVIII). Madrid 1932. pp.717. [2000.]

Carbon dioxide.

LIST of references on carbon dioxide in its relation to the development and hatching of the chick embryo. Library of Congress: Washington 1913. ff.3. [24.]*

Conodonts.

GRACE B[RUCE] HOLMES, A bibliography of the conodonts, with descriptions cf early mississippian species. Washington 1928. pp.38. [500.]

SAMUEL P. ELLISON, Annotated bibliography and index, of conodonts. University of Texas: Publication (no.6210): Austin 1962. pp.128. [1050.]

Curare.

CURARE. Bibliografia. Instituto brasileiro de bibliografía e documentação: Rio de Janeiro 1957. pp.387. [2956.]

Cuvier, Georges.

HENRI DEHÉRAIN, Catalogue des manuscrits du fonds Cuvier (travaux et correspondance scientifiques) conservés à la bibliothèque de l'Institut de France. Paris [vol.ii: Hendaye] 1908–1922. pp. [iii].154+[iii].xii.76. [2500.]

Special Subjects

Darwin, Charles Robert.

J. W. SPENGEL, Die darwinische theorie. Verzeichniss der über dieselbe in Deutschland, England, America, Frankreich, Italien, Holland, Belgien und den skandinavischen reichen erschienenen schriften und aufsätze.... Zweite, vermehrte auflage. Berlin 1872. pp.[iii].36. [750.]

[W. G. RIDEWOOD], Memorials of Charles Darwin. A collection of manuscripts... books and... specimens to commemorate the centenary of his birth and the fiftieth anniversary of the publication of 'The origin of species'. British museum (Natural history): Special guide (no.4): 1909. pp.[v]. 50. [85.]

A. E. S. and J. C. S., Darwin centenary. The portraits, prints and writings of Charles Darwin exhibited at Christ's college. Cambridge 1909. pp.vi.47. [257.]

[R. V. C. BAILEY and J. S. GOSSE], Handlist of Darwin papers at the university library. Cambridge 1960. pp.72. [2000.]

Dew ponds.

DEW PONDS. Science library: Bibliographica, series (no.136): 1934. ff.[i].2. [46.]*

Dismal swamp.

LIST of references on the Dismal swamp. Library of Congress: Washington 1914. ff.2. [24.]*

LIST of references on Dismal Swamp canal. Library of Congress: [Washington] 1926. ff.5. [40.]*

Eugenics.

SELECT list of references on eugenics. Library of Congress: Washington 1911. ff.9. [90.]*
— [supplement]. 1923. ff.7. [85.]*

EUGENICS. Russell Sage foundation: Library: Bulletin (no.3): New York 1914. pp.[4]. [75.]

SAMUEL J. HOLMES, A bibliography of eugenics. University of California publications in zoology (vol.xxv): Berkeley, Cal. 1924. pp.[v].514. [9500.]

BIBLIOGRAPHIA eugenica. Eugenial news: Supplement: Cold Spring Harbor, L. I.
    i. 1927–1930. Editor: Mabel L[avinia] Earle. pp.[ii].349. [3961.]
    ii. 1931–1934. pp.[ii].230. [2735.]
no more published.

EUGENICS. A list of some of the standard and recent books. National book council: Bibliography (no.115): 1929. pp.[2]. [40.]
— Second edition. 1940. pp.[2]. [50.]

Special Subjects

Evolution. [*see also* **Darwin, C. R.**]

ZOOLOGISCHER jahresbericht. . . . Allgemeine entwicklungslehre. Zoologische station zu Neapel: Berlin.

    1886. Referent: Paul Mayer. pp.18. [100.]
    1887. pp.13. [75.]
    1888. pp.17. [100.]
    1889. pp.14. [75.]

amalgamated with the section on general biology.

ANATOMISCHE hefte. . . . Ergebnisse der anatomie und entwickelungsgeschichte. Wiesbaden 1891 &c.

details of this work are entered under Anatomy, above.

HENRY CHURCHILL KING, A selected bibliography of evolution. Oberlin college: Library bulletin (vol.i, no.4): Oberlin, Ohio, 1899. pp.15. [250.]

SELECT list of references on the theory of evolution. Library of Congress: Washington 1910. ff.6. [33.]★

— Additional references. 1915. single leaf. [10.]★

[SIR] J[OHN] ARTHUR THOMSON, What to read on evolution. Public libraries: Leeds 1928. pp.31. [150.]

EVOLUTION. Surrey county library: Book list (no.10): [Esher 1958]. pp.[iv].12. [75.]*

Fabre, Jean.

[J. FABRE], Titres et travaux scientifiques du d^r Fabre [1901]. pp.48. [20.]

Fertility.

ELDON MOORE, *ed.* A bibliography of differential fertility, in english, french, and german. International union for the scientific study of population: Edinburgh 1933. pp.vi.97. [1000.]

Genetics. [*see also* Botany: Genetics.]

ANIMAL breeding abstracts. Imperial [Commonwealth] bureau of animal [breeding and] genetics: Edinburgh:

    i. 1933–1934. pp.400. [2000.]
    ii. 1934. pp.406. [2500.]
    iii. 1935. pp.484. [3000.]
    iv. 1936. pp.519. [3000.]
    v. 1937. pp.492. [3000.]
    vi. 1938. pp.368. [2500.]
    vii. 1939. pp.[iii].520. [3000.]
    viii. 1940. pp.[ii].458. [3500.]
    ix. 1941. pp.[iii].374. [3000.]
    x. 1942. pp.[iii].290. [2000.]
    xi. 1943. pp.[iv].279. [2000.]

xii. 1944. pp.[ii].240. [1600.]

xiii. 1945. pp.[ii].252. [1500.]

xiv. 1946. pp.[iv].286. [1500.]

xv. 1947. pp.[iv].333. [2000.]

— Subject index to vols.8–15. 1952. pp.[v].
162.

xvi. 1948. pp.[iv].428. [1705.]

xvii. 1949. pp.[iv].488. [1698.]

xviii. 1950. pp.[iv].521. [1640.]

xix. 1951. pp.[iv].599. [2023.]

xx. 1952. pp.[iv].476. [1998.]

xxi. 1953. pp.[iv].464. [2019.]

xxii. 1954. pp.460. [1885.]

xxiii. 1955. pp.[iv].504. [2020.]

xxiv. 1956. pp.[iv].484. [2009.]

xxv. 1957. pp.[iv].514. [2185.]

xxvi. 1958. pp.[iv].524. [2287.]

xxvii. 1959. pp.[iv].560. [2196.]

xxviii. 1960. pp.[iv].546. [2366.]

xxix. 1961. pp.[iv].573. [2469.]

xxx. 1962. pp.[iv].682. [3007.]

xxxi. 1963. pp.[iv].648. [3354.]

*in progress; no indexes were published with vols.
iv–vi.*

J[OHN] H[ENRY] KENNETH, Gestation periods. A
table and bibliography. Imperial bureau of animal
breeding and genetics: Edinburgh 1943. pp.23.
[341.]

A[LFRED] ERNST, Bibliographia genetica helvetica 1929–1944. Zürich 1945. pp.[i].504–530. [400.]

MARJORY H. WRIGHT, Viral genetics. A bibliography . . . 1955–1959. National library of medicine: Washington 1960. pp.[iii].34. [309.]*

EUGENE GARFIELD and IRVING H. SHER, *edd.* Genetics citation index . . . with special emphasis on human genetics. Institute for scientific information: Philadelphia [1963]. pp.xxix.864. [375,000.]*

Geotropism.

MARIE CHRISTIANSEN, Bibliographie des geotropismus, 1672–1916. Institut für allgemeine botanik: Mitteilungen (vol.ii = Jahrbuch der hamburgischen wissenschaftlichen anstalten, vol. xxxiv, 3. beiheft): Hamburg 1917. pp.[ii].118. [171.]

supplements formed part of vols.ii–iii of the Mitteilungen.

Germination.

THE EFFECT of light on germination. Science library: Bibliographical series (no.700): 1950. ff.4. [66.]*

— Supplement. . . . (no.786): [1963]. pp.7. [150.]*

Special Subjects

Gibberellin.

GIBBERELLIN. Public library of South Australia: Research service: [Adelaide] 1958. ff.5. [47.]

Herbals.

AGNES ARBER, Herbals, their origin and evolution ... 1470–1670. Cambridge 1912. pp.xviii.253. [150.]

— — New edition. 1938. pp.xxiv.328. [150.]

HORACE MALLINSON BARLOW, Old english herbals, 1525–1640. 1913. pp.42. [40.]

W. L. SCHREIBER, Die kräuterbücher des XV. und XVI. jahrhunderts. München 1924. pp.lxiv. [125.]
published with a facsimile of the Hortus sanitatis, *Mainz 1485; 300 copies printed.*

ALFRED SCHMID, Ueber alte kräuterbücher. Bern &c. 1939. pp.[vi].76. [100.]

JUAN CARLOS AHUMADA, Herbarios médicos primitivos. Buenos Aires 1940. pp.63. [38.]
a catalogue of the author's collection; 250 copies privately printed.

RUTH B. FISCH and MAX HAROLD FISCH, The Marshall collection of herbals in the Cleveland medical library. [Baltimore] 1947.

Special Subjects

Heredity.

ALFRED HENRY HUTH, An index to books and papers on marriage between near kin. 1879. pp. [ii].23. [172.]

LIST of references on the influence of heredity and environment. Library of Congress: Washington 1914. ff.2. [17.]*
— Additional references. 1927. single leaf. [9.]*

LUCIEN HOWE, A bibliography of hereditary eye defects. Carnegie institution of Washington: Eugenics record office: Bulletin (no.21): Cold Spring Harbor, N.Y. 1921. pp.45. [1000.]

GEORGES POYER, Bibliographie sur les problèmes généraux de l'hérédité psychologique. Université de Paris: Faculté des lettres: 1921. pp.15. [250.]

FRIEDRICH OEHLKERS, Erblichkeitsforschung an pflanzen. Ein abriss ihrer entwicklung in den letzten 15 jahren. Wissenschaftliche forschungsberichte: Naturwissenschaftliche reihe (vol.xviii): Dresden &c. 1927. pp.viii.204. [400.]

REFERENCES to the inheritance of coat-colour in rabbits. Science library: Bibliographical series (no.51): [1952]. ff.[i].3. [32.]*

459

Iodine.

IODINE in the animal world. (Farm animals). Science library: Bibliographical series (no.345): 1937. ff.11. [149.]*

—— II. The iodine content of foodstuffs, 1925–37. . . . (no.382): 1938. ff.13. [260.]*

Ivory.

IVORY: a selected list of references. Library of Congress: Washington 1919. ff.4. [44.]*

—— [supplement]. 1946. ff.8. [83.]*

Jardin des plantes, Paris.

LOUIS DENISE, Bibliographie historique & iconographique du Jardin des plantes. Bibliothèque du vieux-Paris: 1903. pp.[iii].268. [728.]

265 copies printed.

Lac.

A. C. CHATTERJEE [A. C. CHAṬṬOPĀDHYĀYA], Bibliography of lac. Calcutta 1933. pp.129. [1250.]

Leather.

FRANCIS D. HORIGAN and CARY R. SAGE, The mycology of leather. A bibliography. QM research & development laboratories: Technical library: Bibliographic series (no.4): Philadelphia [1948]. ff.[iii].37.viii. [115.]*

Leconte, John Lawrence.

SAMUEL HENSHAW, The entomological writings of John L. Leconte. . . . Edited by George Dimmock. Dimmock's special bibliography (no.1): Cambridge, Mass. 1878. pp.[ii].11. [152.]

Life.

V[ERA] A[LEKSANDROVNA] NASEDKINA and G[ALINA] N[IKOLAEVNA] BELAVENTSEVA, Возникновение и развитие жизни на земле . . . Издание 2-е. Государственная . . . библиотека СССР имени В. И. Ленина: Москва 1955. pp.55. [100.]

Liguria.

ARTURO ISSEL, Bibliografia scientifica della Liguria. I. Geologia, paleontologia, mineralogia, geografia, meteorologia, etnografia, paletnologia e scienze affini. Genova 1887. pp.113. [458.]
no more published.

Linné, Carl von.

ORBIS eruditi judicium de Caroli Linnæi . . . scriptis. [Stockholm 1741]. pp.[16]. [60.]

RICHARD PULTENEY, A general view of the writings of Linnæus. 1781. pp.vi.426. [200.]
— — Revue générale des écrits de Linné. . . .

Traduit . . . par L. A. [Aubin Louis] Millin de Grandmaison: avec . . . des additions. 1789. pp. [iii].vi.386+[iii].400. [250.]

—— Second edition. 1805. pp.xv.596. [300.]

LINNAEANA in Nederland aanwezig. Tentoongesteld . . . in het Koninklijk zoölogisch genootschap 'natura artis magistra'. Amsterdam 1878. pp.59.2. [200.]

EWALD ÄHRLING, Några af de i Sverige befintliga Linnéanska handskrifterna, kritiskt skårskådade. Lund 1878. pp.[ii].28. [50.]

WILHELM JUNK, Bibliographia linnæana. Berlin [1902]. pp.10. [188.]

J[OHAN] M[ARKUS] HULTH, Bibliographia linnaeana. Matériaux pour servir à une bibliographie Linnéenne. Partie 1 — livraison 1. Kungl. vetenskaps societeten: Upsala 1907. pp.[iv].170. [1000.] *no more published.*

[ALFRED BARTON RENDLE], Memorials of Linnæus. A collection of portraits, manuscripts, specimans and books exhibited to commemorate the bi-centenary of his birth. British museum (natural history): Special guides (no.3): 1907. pp.16. [30.]

[BERNARD BARHAM WOODWARD and W. R. WILSON], A catalogue of the works of Linnæus (and publications more immediately relating thereto) preserved in the libraries of the British museum. 1907. pp.27. [500.]

— — Second edition. [By Basil H. Soulsby]. 1933. pp.xi.246.68. [4500.]

— — — An index to the authors (other than Linnæus) mentioned in the Catalogue. [By Charles Davies Sherborn]. 1936. pp.59. [3000.]

WALTER T[ENNYSON] SWINGLE, Chronologic list of the dissertations of Charles Linnaeus, 1743 to 1776. With reference to the libraries in the United States containing original editions. Department of agriculture: Washington 1923. ff.[ii].6.[54.] [186.]*

FELIX BRYK, Bibliotheca linnæana. I. Die schwedische Linnéliteratur seit 1907. [II. Neue oder wenig bekannte Linnéoriginale]. Stockholm 1923. pp.52+30. [111.]
100 copies printed.

BIBLIOTHECA linnæana — works by or relating to Carolus Linnæus, his predecessors, contemporaries and pupils, with sequels — from the library of Emil Lindell. Växjö 1932. pp.123. [2150.]

SPENCER SAVAGE, Synopsis of the annotations by Linnaeus and contemporaries in his library of printed books. Catalogue of the manuscripts in the library of the Linnean society (part iii): 1940. pp.20. [300.]

THOMAS R. BUCKMAN, A catalog of an exhibition commemorating the 250th anniversary of the birth of Carolus Linnaeus . . . and the 200th anniversary of the issue of the Systema naturae, 10th edition, 1758–1958. Kansas university: Libraries: Lawrence 1957. pp.50. [73.]

Lintner, Joseph Albert.

EPHRAIM PORTER FELT, Memorial of the life and entomologic work of Joseph Albert Lintner. New York state museum: Bulletin (vol.v, no.24): Albany 1899. pp.[ii].303–400. [1250.]

Loing, river.

PIERRE DOIGNON, Répertoire bibliographique et analytique du massif de Fontainebleau et de la basse vallée du Loing. Travaux historiques et scientifiques. Association des naturalistes de la vallée du Loing et du massif de Fontainebleau: Fontainebleau 1958. pp.56. [3500.]

Meek, Fielding Bradford.

JOHN BELKNAP MARCOU, Bibliography of publications relating to the collection of fossil invertebrates . . . including complete lists of the writings of Fielding B. Meek. United States national museum: Bulletin (no.30): Washington 1885. pp.333. [Meek: 106.]

Miscegenation.

MAGNUS MÖRNER, El mestizaje en la historia de Ibero-America. Estocolmo 1960. ff.59.*

Morphology.

WILTON M[ARION] KROGMAN, A bibliography of human morphology, 1914–1939. University of Chicago: Publications in anthropology: Physical anthropology series: Chicago 1941. pp.xxxi.385. [10,000.]*

JUAN COMAS [CAMPS], Bibliografía morfologica humana de América del sur. Instituto indigenisto interamericano: México 1948. pp.xxiv.208+8 maps. [2971.]

СОВЕТСКОЕ медицинское реферативное обозрение. Нормальная и патологическая морфология с эмбриологией. Москва.

    i. Редактор А. И. Струков. 1949. pp.144. [600.]

in progress?

Nitrofurazone.

NITROFURAN. Biologic bibliography 1957. Eaton laboratories: [Norwich, N.Y. 1958]. pp.3–105. [1900.]*

Special Subjects

Oxygen.

LIST of references on the influence of oxygen and air with excessive oxygen on living organisms. Library of Congress: Washington 1915. ff.2. [17.]*

Photosynthesis.

ФОТОСИНТЕЗ. Указатель отечественной иностранной литературы. Московский... университет: Научная библиотека им. А. М. Горького: Научно-библиографический отдел: [Москва].

    i. 1951–1958. [By] Н. В. Арциховская.
      1961. pp.387.403.507. [8995.]
in progress.

Plumage.

THE EFFECT of ovarian and testicular hormones on plumage. Science library: Bibliographical series (no.322): 1937. ff.3. [57.]*

Pollen.

M[ARK] I[LICH] NEISHTADT, Спорово-пыльцевой метод в СССР. История и библиография. Академия наук СССР: Институт географии: Москва 1952. pp.223. [926.]

Special Subjects

Rauwolfia.

BIBLIOGRAPHY on rauwolfias. Pakistan national scientific and technical documentation centre: Pansdoc bibliography (no.100): Karachi 1959. pp. [viii].79–196. [3000.]

Reclamation. [*see also* **Conservation.**]

JOHN J[AMES] GAUL, Reclamation 1902–1938: a supplemental bibliography. Public library: Regional checklist (no.6): Denver 1939. pp.[ii]. xxvi.98. [1350.]★

Reefs.

W[ILLIAM] E[MERSON] PUGH, Bibliography of organic reefs, bioherms, and biostromes. Seismograph service corporation: Tulsa [1950]. pp.xxxi. 139. [1750.]

Roots.

ROBERT H[AROLD] MILLER, A selected reference list on the morphology and anatomy of roots. SL bibliography series: New York [1960]. pp.[iii].21. [450.]★

Rotenone. [*see also* **Derris.**]

RECENT literature on the toxicity and insecticidal value of rotenone. Science library: Bibliographical

467

series (no.135): 1934. ff.[i].4. [71.]★
— Supplement . . . (no.240): 1936. ff.8. [146.]★

Sex, sexual system.

VIERTELJAHRESSCHRIFT über die gesamtleistungen auf dem gebiete der krankheiten des harn- und sexual-apparates. Berlin.
    i. 1896. pp.516. [1450.]
    ii. 1897. pp.788.24. [2250.]
    iii. 1898. pp.760. [2400.]

LIST of references on the sex problem and sex instruction. Library of Congress: [Washington] 1919. ff.3. [32.]★

ROGER GOODLAND, A bibliography of sex rites and customs. 1931. pp.v.752. [9000.]

CHRISTIAN sex education. List of pamphlets. British council of churches: 1946. pp.[4]. [30.]

BIBLIOGRAPHY of sex guidance and human relationships. Society for sex education and guidance: [1952]. pp.44. [400.]

D[ERRICK] S[HERWIN] BAILEY, The theology of sex and marriage. A short guide for readers and students. Church of England: Moral welfare council: 1953. pp.28. [175.]

G[EORGY] Z[AKHAROVICH] INASARIDZE, Биб-

Special Subjects

лиографический указатель отечественной урологии и смежных областей за 100 лет, 1855–1955. Тбилиси 1959–1962. pp.438+ +620. [20,678.]

BIBLIOGRAPHY of reproduction. A classified monthly title list compiled from the world's research literature. Vertebrates including man. Reproduction research information service: Cambridge.★
   i–ii. 1963. pp.xx.932.128. [8316.]
in progress.

Sterility.

WILLIAM ORR and F. FRASER DARLING, The physiological and genetical aspects of sterility in domesticated animals. . . . With a bibliography by miss M. V. Cytovich. Imperial bureau of animal genetics: Edinburgh 1932. pp.80. [550.]

Surface tension.

FERDINAND HERČÍK, Oberflächenspannung in der biologie und medizin. Wissenschaftliche forschungsberichte: Naturwissenschaftliche reihe (vol.xxxii): Dresden &c. 1934. pp.xii.220. [750.]

Tephrosia. [*see also* Derris.]

R[URIC] C[REEGAN] ROARK, Tephrosia as an

insecticide — a review of the literature. Bureau
of entomology and plant quarantine: Division of
insecticide investigations: Washington 1937. pp.
165. [601.]★

Weeds.

WEED abstracts. Commonwealth agricultural
bureaux.

    xi. 1962. pp.viii.443. [2022.]
    xii. 1963. pp.viii.   . [1908.]
in progress; vols.i–x were issued by various organi-
zations in duplicated form.

SOME references to the use of calcium cyanamide
as a weed killer (1932–1958). Science library: Bibliographical series (no.765): [1958]. pp.4. [100.]★

Yellowstone national park.

CARL P[ARCHER] RUSSELL, A concise history of
scientists and scientific investigations in Yellow-
stone national park. With a bibliography of the
results of research and travel in the park area.
Office of national parks: [*s.l.* 1934]. pp.144. [1600.]

Yosemite.

FRANCIS P[ELOUBET] FARQUHAR, Yosemite, the
big trees and the high Sierra. A selective bibliography. Berkeley 1948. pp.xii.104. [150.]

Special Subjects

Zein.

DOROTHY M. RATHMANN, Zein. An annotated bibliography, 1891–1953. Mellon institute: Bibliographic series: Bulletin (no.7): Pittsburgh 1954. pp.vi.118. [900.]